Practical Guide to Business Continuity Assurance

For a listing of recent titles in the *Artech House Technology Management and Professional Development Library*, turn to the back of this book.

Practical Guide to Business Continuity Assurance

Andrew McCrackan

Artech House
Boston • London
www.artechhouse.com

Library of Congress Cataloging-in-Publication Data
A catalog record for this book is available from the U.S. Library of Congress.

British Library Cataloguing in Publication Data
McCrackan, Andrew
   Practical guide to business continuity assurance. — (Artech House technology management
library)
   1. Crisis management   2. Risk management   3. Strategic planning
   I. Title
   658.4'056

   ISBN 1-58053-927-0

Cover design by Gary Ragaglia

© 2005 ARTECH HOUSE, INC.
685 Canton Street
Norwood, MA 02062

International Standard Book Number: 1-58053-927-0

10 9 8 7 6 5 4 3 2 1

To Claire-Louise, for your love, support and tolerance.

Contents

1

Introduction

In business we are faced with many threats to the continuance of our trade. The usual "acts of God" and the like are of continued concern as is the newly elevated threat of terror and other prevailing security-related risks. There is much misunderstanding of Business Continuity Management (BCM) and as yet no commonly practiced methodology for the assessment of an organization in this regard. Nor is there structured implementation of capabilities to directly address the risks presented to us in the modern world. These newly emphasized risks in partnership with the inherent vulnerabilities resulting from the lack of a structured approach to BCM calls for a more robust and measurable means of protecting our organizations. This in turn gives rise to a new era of continuity management that is continuity assurance. As the name implies, continuity assurance is concerned with actively planning to avert the threat or reduce its effect rather than simply planning how to recover from the act, though this recovery is implicit in the approach detailed in this text.

The crux of this approach is the integration of physical and logical security with disaster recovery, health and safety, and general business continuity and risk management functions within the organization.

The methodology set out within this work will provide the reader with a blueprint with which to assess their organization with regard to continuity assurance capability. It may further be utilized to determine what needs to be done to increase their organization's capability and provides guidelines and advice on how best to implement such corrective measures in a detailed, practical, and pragmatic fashion.

Furthermore, by providing such a business continuity framework it is hoped that a benchmark can be accepted, in the form of the continuity assurance achievement rating (CAAR), with which to measure organizations against each other, this may be used to demonstrate to peers and other industry bodies a sound capability in preventing or managing terrorist attacks or other general continuity-impacting situations. The implementation of this standard to a given quantitatively measurable level may be further utilized as a leverage point against which corporate insurance premiums against threat and disaster may be negotiated and potentially reduced.

1.1   Purpose and Scope

The purpose of this work is to document and distribute a blueprint that can be used by any organization of any size to gauge their continuity capabilities. Furthermore, the method outlined within may be used to analyze gaps between what is currently employed in your organization and what is ideal. This knowledge combined with the instructions and guidelines documented herein may be used to advance your organization through the seven defined levels of continuity assurance to a superior position of continuity capability.

This work documents the following:

1. A cursory review of current business continuity and disaster recovery strategies;
2. Analysis of shortcomings of traditional approaches in a modern world;
3. Information on new factors that influence the way we ensure continuity;
4. An approach to incorporating these new factors and added efficiencies in a new business continuity method;
5. A full end-to-end methodology for continuity assurance;
6. Advice on using this methodology to determine strategy for your organization;
7. A detailed rating system to provide criteria to implement against, and a system for assessing your achievements;
8. A seven-level approach for developing best practice continuity assurance;
9. Instructions on how to manage the implementation of this methodology and system;
10. Views on keeping up with new developments and evolving continuity assurance within your organization.

1.2 How to Use This Book

While compiling this work I have attempted to keep the language informal and conduct the relationship between you, the reader, and myself as I would if dealing with you personally in a consultative role advising to your organization. This serves to connect you personally to the outcomes and hopefully enables you to relate the work directly to your own experiences within your business.

The language used is, by my own admission, not the simplest and is intentional in an effort to reduce the mass of this work to a book that is short and concise. I do not favor books that labor the point and are repetitive in their message. This book is intended as an information source and practical guide. It is not intended as a literary work. You may need to read some sections more than once, or revisit earlier sections with the benefit of the knowledge learned in latter sections.

This book may well be read from front to back but is designed so that each chapter stands alone and addresses completely the chapter topic. In this way the book may be used as a reference guide. It is my intention that you have this work at hand and may refer to it topic by topic as and when inclined.

The information contained herein may be built upon and refined over time. To this end I have established a web site that will detail the latest thinking on continuity assurance and may additionally be utilized as a continuity assurance forum. The URL is: http://www.continuityassurance.com.

1.3 Current Standards and Literature

Traditional, or rather current, BCM approaches are well respected and are fundamentally sound albeit lacking in some areas of detail and practical applicability.

Current standards for business continuity are contained within documents that are principally concerned with information security management. The relevant ISO standard is ISO17799. The parts that relate to business continuity are reasonably low in detail at this time. There are many efforts underway to address this situation. One such effort is embodied in the PAS56 (publicly available specification) guide to BCM, which has been produced in cooperation with The Business Continuity Institute (BCI).

The idea behind this work is to provide a more detailed and usable framework for BCM. Many attempts to accomplish this by others have resulted in largely academic and theoretical works which are difficult to disseminate and require a great deal of interpretation and further analysis to be of practical application. My emphasis and my goal in this work is to provide a practical and pragmatic means of implementing better business continuity capabilities within your organization.

I favor an integrated approach to BCM. It is the implementation of business continuity across an organization rather than just the creation of a special set of rules and roles that come into play only in a crisis situation. As the saying goes, "The devil is in the detail." My aim is to get far enough into the detailed planning process to get you over the most difficult areas of implementing a functional business continuity capability. This book will provide you with a complete end-to-end methodology for implementing business continuity, and show you how to use it.

All this does not diminish the value of the work done by other experts, institutions and companies in this field. I prefer to see this work, although it is an advancement on traditional methods, as complementary to the existing body of knowledge in BCM.

There is not a great volume of quality literature available that is recent and deals specifically with BCM. I have detailed some titles for further reading and referenced them at the end of this introduction [1–3]. I also have a collection of books on related topics that people always seem to find very useful [4–5].

1.4   Business Continuity and Disaster Recovery

Many people are confused about the difference between business continuity and disaster recovery; indeed many use the terms interchangeably. Business continuity is concerned, as the name implies, with the continuance of business in the face of any unusual or unforeseen event. A "disaster" is one such occurrence of an unplanned event. There is no standard definition of the term disaster but it is usually an event that has caused significant damage to business operations and requires recovery from. This means that the disaster has exceeded the tolerances put in place to withstand localized fault, failure, or destruction of business infrastructure. A disaster will completely stop single or multiple business processes requiring the organization to execute some level of specific recovery plan, depending on the exact nature of the disaster. In this way disaster recovery is a subset of BCM, as is contingency planning, high availability planning, and the like.

Traditional business continuity management is really concerned with the requirements side of continuity planning. The business continuity professional will analyze an organization to identify the critical business processes, critical staff, potential risks, and threats to the continuity of the business and the tolerance of the customer and business to the unavailability of those business processes. These requirements are distributed to the relevant sections of the organization that need to address the issues raised, and put in place mechanisms or plans that can be used if an event occurred that impacted the continuity of the business. For the great majority of organizations this task will fall to the information technology (IT) department, as IT facilitates the majority of key business processes in the modern company. Have you ever heard the phrase "The computer won't let me do that" or "I can't serve you because the system is down." Information technology has become a strong management tool and a way of ensuring that anything that is done in a company that has a cost associated is tracked for auditability and control reasons. The down side of this is that companies are reliant on IT systems to be able to go about even the simplest of operations.

So given that IT professionals have already analyzed the workings of the business, and in the majority of cases control the workflow through the business side of the organization, why is generic BCM generally disconnected from the IT–implementation side of the continuity equation? The reason is typically resource related. Business continuity professionals are not always IT professionals. There is a skills gap that separates the requirements gathering and strategy from the implementation. We will discuss this at some length later.

Disaster recovery is the implementation of a response capability to a specific type of event that impacts the continuity of the business. Business Continuity Management is responsible for the overall identification of potential events, the likelihood of the occurrence of the event, and the predicted impact on the organization. Business continuity management puts in place plans to deal with such occurrences. Disaster recovery is essentially a plan, with supporting infrastructure, which is enacted in the event of a disaster.

Most companies treat disaster recovery as an add-on to existing operational infrastructure and processes. It's a function that is bolted on to the side of the company, tested occasionally but rarely given too much emphasis. The principle reason for this is, as with all things, a matter of finance. A company will assess the cost of downtime verses the cost of systems resilience. There is usually a compromise struck that results in a level of acceptable downtime of operations at an acceptable cost, or loss.

There are many factors to take into consideration when assessing impact of a disastrous event. They are generally grouped under the following headings, referred to as the continuity indicators:

- Financial;
- Reputation;
- Customer service;
- National security;
- Health and safety;
- Regulatory.

Business continuity management acts to safeguard the continuity of business. Its job is to make sure that the business continues to operate at all times, no matter what event has taken place to interfere with this. In this way BCM is a form of quality assurance (QA), or at least is similar to the way QA operates throughout the organization.

1.5 Gap Analysis

I have been alluding to a fairly clear gap between current thinking in business continuity circles and my own thinking. The gap can be summarized in these few bullets:

- Lack of knowledge transfer between the spheres of puritan business continuity and technical disaster recovery;
- IT security and physical security autonomous to each other;
- No clear quantitative methodology in place to rate and benchmark;
- Lack of integration of health and safety with the business;
- Continuity planning is isolated, not integrated.

All of these issues can be reduced to a single root cause, which is: lack of integration between the business, IT, security, and health and safety.

Note that when I refer to the "business" I am referring to the areas of the organization that are concerned purely with revenue generating activities. In IT it is common to refer to these parts of the organization, which are the customers of the IT department and other support functions, collectively as the business.

Without an integrated approach combining these four organizational facets it is not possible to execute a coordinated response to a crisis situation. Lack of integration will mean that your response is likely to be segmented over the response timeline leading to gaps in continuity or quality of response.

For example, technology may ensure that systems are recovered in short time but if the people that utilize these systems are not able to get to a point of access then the underlying business processes that they support will fail to restart. If some of these people are key people either in the business process or the recovery process and one is injured in the crisis, say by falling down an inadequate building escape, then there is a possibly unforeseen gap in knowledge that may cause a breakdown in the recovery plan. Likewise if the crisis is more general and not specific to your organization you may be faced with people rushing back to their families to provide for their safety.

The last point here we will talk about in more detail later but be aware that in a crisis situation you may need to provide adequate assurances for the people that make up your organization. These assurances may be in the form of measures you will need to take to safeguard them and their families in the event of such a crisis to ensure that they remain focused on the continuity of your business. It will almost certainly be more cost effective for you to provide for this than try to work around the absence of a key person in your business or recovery process.

1.6 Integration of the Organization

As previously stated, a principle problem with the way business continuity is approached in many organizations today is the lack of an integrated approach to business continuity stemming from a lack of integration within the organization itself. Of particular concern is the lack of integration between business, IT, security, and health and safety.

In the world today the core of many business functions is IT. Without it plants do not operate and people are not billed for services. In some cases services cannot be provisioned or provided at all.

It is always of some amazement to find that technical functions in companies still reside in an IT department or similar segregated section of the organization. In this day and age most business relies on IT as its core enabler. Without IT most companies would not function at all. Even in the most IT intensive sectors of banking and telecommunications the segregation can be found. We commonly refer to the IT and the business to the degree where relationship managers are often found sitting between the two sections to control the interface between the technical and nontechnical. This structure is an inefficient one. It exists in principle for reasons of business focus and staffing skill issues but this landscape is changing. With these changes in staff IT knowledge and business IT convergence will come a new more efficient organizational model. Implicit in this model will be its ability to combine several logical functional areas that are complementary and until now partly duplicated in both the business and IT sections of an organization.

1.7 Business and IT

The lack of integration between the business and IT sections essentially comes from a legacy knowledge gap between these two areas of the organization. In the case of many businesses it has been possible for the IT staff, business analysts, and the like, to pick up the detailed workings of the business. For the business resource the detailed workings of the technology that supported their business was, and in many cases still is, very much a mystery.

This is a particular problem in those businesses that are essentially IT based. Although most business today is IT reliant there are some whose business *is* IT but they have not realized it yet. Banking and telecommunications are good examples but there are many others. Many of these organizations, but not all, are still segmented along the IT boundary. This means that there are the users of the technology on the one side and the creators on the other. A relative plethora of people sit in-between acting as translators. These people generally sit within the IT organization and take the form of business analysts, line managers, project managers, and relationship managers or the like. In some organizations the business will also put up an interface in the form of a relationship manager or, as is often the case, duplicate some of the IT support or business analysis functions within their organization in order to gain some additional control and ownership over the technology.

So what is the problem with this? Well, it is not very efficient for one, as functions are often duplicated, and the disconnects that often occur between the two organizations lead to iterative rework loops. These loops occur when the business defines a need that may be a need but may also simply be a want. IT attempts to translate this requirement into something real and tangible or, in many cases, provides an alternate solution to the requirement using the existing technical infrastructure. In the event of new technology needing to be developed to address the requirement the IT organization will produce something based on what the business has asked for and supply it to them for acceptance. This is usually where the business realizes that it has either asked for the wrong thing or did not need it anyway. This may seem a little harsh and no, it is not the way it is meant to work. The business analyst is there to make sure that situations like this do not develop. That is all well and good in an ideal world but the reality is that this disconnect will always occur in an organization where there is an underlying disconnect between the business and IT to begin with. You only need look at the number of new software development projects that fail to know that this is true.

Why does this happen? If the organization is segmented and disconnected the underlying problem will almost certainly be an understanding gap along the IT boundary. This lack of understanding leads to fear. Fear breeds mistrust between the business and IT organizations. This also creates rivalry and pride. Pride leads to self-righteousness, which leads to the following scenario. The business defines a requirement that may or may not be sensible, but it is what they want. IT is bound by a formal service agreement, which they have put in place because they do not trust each other. This contract between the business and IT creates a culture where neither party is necessarily working for the best interests of the organization as a whole. They each tend to lose sight of the big picture in their endeavors to keep the other party in control. So IT blindly does whatever the business has asked for. Yes, if they do not agree they may push back a little but the business will win in the end, after all, they are the customer. So the new technology is developed and delivered to the business. The business will look at it and will more than likely do one of the following:

1. Tell IT that the requirement was misunderstood and that was not what they had asked for.
2. Reject the technology as not working.
3. Realize their error but not wanting to look foolish, accept the technology and then never actually implement it, or, implement it but never use it.
4. Realize their error and attempt to negotiate a way forward with IT.
5. Find any other way possible to save face.

The most likely is of course outcome number two, which sends the business and IT into an often never-ending rework loop. This loop continues until the project or work package runs out of budget and is eventually terminated with the business placing most, if not all, of the blame on the IT organization's inability to deliver the requirement.

Of course there are also situations where IT is not able to get it right technically but in my experience these situations are also generally arrived at because of some level of business error in understanding. IT will not usually be blameless in these situations though. An example of this may be the business selecting an off-the-shelf software package to fulfill a set of requirements or simply to keep up with the technology their competitors are using. Or perhaps the software vendor has done a particularly good sales job and convinced the business that they need this new software package. The IT department will then get stuck with implementing something that may not meet the business requirements at all. Worse still the business may require IT to modify or customize the package to fit within their business operations. This may simply not be possible, let alone cost effective, regardless of what the software vendor has promised. The business may well have asked the opinion of IT at some stage in this process but this does not mean that they will have listened to or understood the advice. IT, in their quest to keep work on the table and thus keep their jobs may agree to things under a level of duress that a reasonable person would not. Such is the way of the world.

Creating a customer-driven culture between segments of an organization can be a good thing but I think this is an example of where the model fails. As in any business, sales and cash flow are the name of the game and often companies will make decisions on factors other than what is sensible technically or sensible for the client's business. The old adage, "Give them what they want, not what they need" runs true. I have found this to be the case when selling product or consulting services. Most people have an idea of what they think they need before they talk to you. It is pretty difficult to sway them from their preconceptions, especially when your competition is agreeing with them and telling them everything that they want to hear. What people think they need is actually what they want. What they really need is often quite a separate thing all together. It is often the case that a budget has been set around what they want. The price of what they need may be much higher. This clearly compounds the problem further. Indeed, it has been said in consulting circles that you can sometimes be too knowledgeable about a particular topic to be able to bid a project based on that topic. The problem being, that you know all the pitfalls and problems and will cost-in dealing with these issues. You may also know that extra effort or resource is required in certain areas. A bidder ignorant of these facts will pitch something far simpler and almost certainly cheaper. Yes, that means that the successful vendor may come unstuck during the project but projects can be renegotiated. There are always options once both the customer and vendor are beyond the point of no return. The key is winning the business in the first place. Unfortunately, when you apply this operating model internally within an organization, between the business and IT, you have entered a very unfavorable place and one that is almost certainly not in the best interests of the organization as a whole.

So what has all this got to do with business continuity? Business continuity is about constructing a solid and resilient organization that can deal with extenuating circumstances or situations. A building may be strong and long lasting and be built from high-quality materials but if it has not been designed to withstand high winds or minor earthquakes then it will not survive for very long. Such is the case with your organization. An organization has to be designed with business continuity in mind. Sure, you can strap bracing structures to your building after the fact but they will never be as strong as having an integrated design that incorporates additional strength to begin with, and they will surely be ugly to look at and difficult to manage, and costly!

We have discussed how having business and IT segmented is a bad thing because of inefficiencies and high costs and poor decision making and so on but that is not all. A segmented organization is dangerous to the continuity of your business. Misunderstandings between the business and IT, whatever form they take, can have disastrous consequences. High rework means a high rate of change in the technical environments. Whether these are production, testing, or development environments they are all environments that are necessary at some level for the continuity of business operations. High rate of change means increased risk of error. The single biggest cause of system outage in any organization is that resulting from a failed change implementation. System outages may mean impact to users and/or customers and impact to revenue or any of the other key business continuity indicators.

1.8 Health and Safety

Health and safety is one of the key business continuity indicators. Any event that impacts health and safety may have significant impact on your business as a whole. In my opinion, this is the most important of the business continuity indicators. People are the principle asset of any business, without them the business does not function. You may have heard of the paperless office but no one has yet been able to cope with a "peopleless" company. Even businesses that are heavily technology driven such as Internet-based business, for example, have people working behind the scenes to make it all happen. If the people that work for you are not safe at work or do not feel safe at work then you will not get the best out of them, and you may suffer interruption to your business as a result of injury or death. Most companies are content with making sure that they meet the regulatory requirements on health and safety set out by the relevant industry or governmental body. Unfortunately, in many parts of the world these regulations are particularly weak. Even in countries where you would expect these regulations to be quite comprehensive they are sometimes not.

Few companies look at health and safety with a view to assessing their workplace independent from the letter of the law. In many cases the company has not come to grips with the risk, and is ignorant of the results the law is attempting to effect. I have been in many buildings where the fire escape is used as the smoking room, endorsed, and often used, by the company executive. One has to see the ridiculousness of this situation. What happens if there is a fire in the fire escape? It is a prime example of a company having a capability but not really understanding why or how to use it appropriately.

Many regulations revolve around fire but fire is not the only health and safety issue. The infrastructure set-up in companies to deal with the event of fire may be used for other purposes. Fire wardens roles could be expanded to encompass that of a general safety officer, or security officer. In some more safety-orientated companies such as those in the mining industry, for example, this is the case. Work places can also be designed with health and safety considerations integrated into the building design. By this I am talking about more than fire doors and fire escapes. Most safety issues at present tend to revolve around accidents or natural disaster. There is an increased need for security considerations to be integrated into safety precautions. In an explosion more people are typically killed by flying glass than by anything else. The addition of simple shatter proofing to glass is enough to prevent this. Note that shatter proofing should be applied to all glass in the workplace, not just windows. Glass partitions are equally deadly.

In a world that is obsessed with the impacts of poor ergonomics, cell phone emissions, passive smoking, effects of fluorescent lights, and all sorts of other things don't you think it is time that someone paid attention to the bigger, more threatening issues. Not that these things are not important, indeed we seem to be moving in a positive direction to a world where employees are better looked after, but it is simple to replace lights or a chair. The problems we are faced with in battling threats such as terror and natural disaster are much more complex to address. People have a right to be safe in their workplace in all respects and companies have a responsibility to do all that they can to ensure this is the case.

The tragic events of September 11, 2001 saw many people lost as they battled to evacuate via fire escapes that would seem impractical and not fit for purpose. To have stairs on such a tall building where it takes more than an hour to traverse them is not a practical means of escape in event of disaster. Furthermore, what about disabled people in wheelchairs, how are they meant to escape in such a situation? Not enough thought has gone into designing workplaces for protection and evacuation. Just like business continuity and disaster recovery is bolted onto the organization's structure, so to is a fire escape bolted onto the side of the building, with little thought of the practicalities of actually using it. In some specific types of workplace environment designers have been called in to look at the areas of safety and evacuation. For example, some oil rigs are fitted with what are called skyscapes. These are like a tubular slide that extends from the platform to the ocean. They are a fast means of escape compared to lowering lifeboats. We have all seen the use of similar slides on airplanes. It is clear that there are people out there that are willing to think up all means of evacuation mechanisms, so why are we still stuck with the stairwell in buildings. The answer is that they are all that is required to meet the regulatory requirements. Industries that operate unusual or high-risk work environments have gone several steps further because of the elevated risk to their staff. However, in a world where terrorism can strike at the heart of any city, and has, can we still consider the average office building a low risk, safe environment?

1.9 Security

The key integration problem with security revolves around the segmentation of physical workplace security and IT security. It stems from the systemic segmentation between the business and IT. Where IT security sits, quite obviously, within IT, the physical security department will sit somewhere in an annex of the organization like building management or something similar. Additionally, most companies outsource their physical security needs to specialist companies. I have only seen one organization that had an integrated security policy where a central security group sat directly underneath the board of the company, but this company was an obvious terrorist/criminal target so they had put a little more thought into things than the average organization, large or small. This company did not use any external agencies for security due to the fairly obvious problems with these sorts of companies. This company also had its own drivers and cars so that they did not have to deal with taxis on the premises.

Regarding these generic security companies, the problem is usually staff related. The generally poor quality of staff recruited into these companies being of low education levels with only minor specific training and the high staff turnover and use of casual labor is a significant problem. As an example, I was once in the Los Angeles airport and watched about eight of these security personnel wrestling with each other around an unused scanning machine. It was like watching kids in a schoolyard. My belief is that these people do little for your overall security. Hire a specialist consultant and have him or her recruit and train security staff directly. Pay them well and look after them. Using these companies is another example of organizations not really understanding what they are about. They know they need security so they get a guard. There is some thinking missing here somewhere. They have not understood the root causes of the problems they are trying to solve. Without this understanding you will be reacting with tactical measures rather than implementing a complete strategic solution. Without full understanding of the problem you may act to worsen, not better, your security situation.

1.10 Terrorism

Direct loss of life is a likely consequence of a terror attack, but further consequential losses, should the effects be widespread across multiple organizations, can be equally severe in terms of general economic impact. Let us not forget that our companies support and feed all of the people employed by them and a percentage of the people working for the companies who are in turn contracted by the company for supplies, services, and the like. If the company suffers service interruptions that cripple its ability to turn a profit then all of those people will be impacted most severely. This of course is one of the aims of the terrorist, to cripple the economy by firstly direct impact and, as a consequence, the mere threat of impact.

Of all results of terror, that with the highest impact to continuity of business is employee perception and panic. People have a right to work in a safe environment and to be free from threat and the psychological effects that the threat of terrorist acts brings.

Recent bombings in Saudi Arabia were aimed at scaring off the expatriate work force there. Although expatriate workers are paid relatively well they are not paid enough to risk their lives. As soon as the terror reaches the point that the average worker feels that he or she is risking his or her life then they will leave. The theory is that if all the expatriates got up and left all at once then the country would be instantly crippled and the terrorists would have an opportunity to oust the current royal family. I have consulted in Saudi Arabia and estimate that if all the British, American, and other Western workers left the country at about the same time the country's core infrastructure would grind to a halt within a matter of weeks. In the West the timeframes are of course much longer but the same scenario applies. Many people did not go back to work in New York for some time after September 11, 2001. The cost of this act alone to the economy was huge.

The problem is therefore twofold. How do we protect our organizations to the degree that our employees feel, and indeed are, safe; and how do we finance what will for most be a greatly improved defense and recovery capability?

How many organizations do you know that are really prepared for a terrorist attack? How well are they prepared for other types of disaster?

1.11 Finance and Motivation

Unfortunately, all things can usually be equalized in financial terms. That is, it is possible to predict how many customers could be lost as a result of an event that impacts customer service. This loss can be estimated in dollars per day, or hour. It is also possible to predict the financial impact on a company due to loss of life. This is where we get into the less savory calculations. In most cases the perception is that the only way that loss of life as a result of a disaster will impact a business financially is via an increase in their insurance premium. Any consequential impact to reputation will usually be negligible as most people make allowances for the fact that it was a disastrous event that caused their service interruption. By this I mean a natural disaster or accident or act of willful damage or the like. This point, and the insurance, is only in doubt where some level of negligence can be proved. This begs the question, in failing to protect against a disaster that could be foreseen, is a company being negligent? Furthermore, in an age where acts of terror strike indiscriminately, is there not a predictable risk to all businesses. Are we not all bound to protect our business and our employees against such events?

1.12 Insuring for Disaster

Many things in this world are driven by insurance. Though it is meant for use in treating effect, it is often instead used in cause. That is, insurance itself drives many behaviors in business. It is often the proverbial tail waging the dog. This is most recently noticeable at airports. Have you ever been in the situation where you have passed all the airport security checks and proceed to the airline gate for boarding and they have yet another group of people that check your passport and go through your bags again. Do you really think that this increases security? Do you really believe that someone working for a subcontracted security firm with little training, on minimum wage and at the end of a 12-hour shift is going to be able to pick up something that has not been detected at previous official checkpoints? Yes, perhaps, but do you think that the likelihood is high enough that the airline deems this necessary, or do you think that they have negotiated a deal with their insurers.

You will be familiar with the sorts of questions insurance companies ask when you are taking out insurance on a car. Most companies will give discounts based on whether you have an alarm or an antitheft device fitted to the car. Some discounts are even proportional to how much you spent on fitting the alarm system. They will also take into account the type of car and how popular it is to steal. They will further reduce your premium if you put the car away in a garage overnight. This method of premium reduction is based on increased security reducing the risk and therefore the insurance company's likelihood of having to pay out on the insurance. A reduction in risk is a reduction in liability for the insurance company. All insurance is cost based on liability. When you take out insurance the insurance company, or their underwriter, is taking on potential debt. They effectively sell that debt back to you in the form of a finance agreement with you paying a monthly charge. However, just because we have insurance does not mean we do not bother locking our cars. In fact, this may negate your insurance entirely.

Business insurance works no differently, the numbers are bigger and the number of variables greater. Indeed, some companies do give premium reductions based on having certain security measures in place but there is no standard currently in place for what is acceptable security and what is not. Most companies have limits on what they will insure for. Beyond a certain level it just becomes too expensive. In any event it is near impossible to insure against business losses due to downtime as a result of any event, whatever the cause. Companies thereafter enter the realm of self-insurance. Where they endeavor to layoff risk by investment in security, continuity, and recovery.

The problem with this is that there is no coordinated or standardized approach across companies and no standardized rating process for insurance companies to benchmark one organization against another, or against a model company.

Companies are also prone to putting in place security or safety measures for the purpose of satisfying regulation. Of course the underlying motivation is not necessarily one of security or safety but of avoiding prosecution, fine or, impacting the insurance premium. A good example of this is again our old friend the fire escape. How many fire escapes have you seen that are actually practical. For buildings built before the regulation was put in place the fire escape is treated much the same as disaster recovery is in that it is literally bolted on to the side of the building. Newer buildings are little better. On September 11, 2001, many employees died while trying to escape via the fire escape. When buildings are erected how much thought does the architect give to the practicality of putting an escape stairwell on the side of a skyscraper. Is this thought about at all or is it just done blindly to meet a regulation without any thought of use. How many different types of fire escapes have you seen? Do we have poles or slides or fireproof elevators or escalators or anything different? I must admit I have never seen anything other than the standard exterior stairwell. Someone invented the slides used on airplanes as a practical means of escape. I have not seen as much ingenuity as this go into an office building. How many companies purpose build or renovate their buildings? When a company employs a disabled person, in a wheelchair say, and they give that person a desk on the 20th floor of their office building, how do they expect that person to escape down the 20 levels of stairs? Is there a regulation for this? There are regulations regarding disabled access but what about egress. The point is that companies are not motivated by what is practical or necessary. They are motivated by what they are made to do by law or are made to do because not to do it would have too much financial consequence. Therefore I believe that what is needed is some motivation or incentive for companies to think about these things and to get the answers right. I do not believe that they deliberately do not do it right; the company has just not noticed the problem. Attention follows the finance and unless some financial or legal incentives are provided for this issue it will not get anyone's attention.

A company does not act to solve a problem until the problem is identified. Companies are typically reactive. They do not go out looking for problems in their organization that need solving. This is not usually deemed cost effective. We need to make it so.

It is the purpose of this book not only to define a method for measuring and rating a company's continuity capabilities but to propose to utilize this new standard, itself the implementation of measures, in a way which finances the implementation of these very measures via the insurance process.

It is intended that if enough companies employ the methods detailed in this book then they may collectively put pressure on their insurers for a reduction in premiums. This, over time, should compensate them for their initial investment in continuity assurance.

References

[1]   Hiles, A., *Business Continuity: Best Practices—World Class Business Continuity Management*, 2nd ed., Brookfield, CT: Rothstein Associates, 2003.

[2]   Barnes, J. C., *A Guide to Business Continuity Planning*, New York: John Wiley & Sons, 2001.

[3]   Von Roessing, R., *Auditing Business Continuity: Global Best Practices*, Brookfield, CT: Rothstein Associates, 2002.

[4]   Carnegie Mellon University, Software Engineering Institute, *The Capability Maturity Model—Guidelines for Improving the Software Process*, Reading, MA: Addison-Wesley, 1994.

[5]   Bentley, C., *PRINCE 2—A Practical Handbook*, Woburn, MA: Butterworth-Heinemann, 1997.

2

Approach

As with any problem we must have an approach to solving it. Before we begin let us first summarize what the key problems are with present business continuity methods and the achievement of business continuity goals in organizations today.

The key issues that we have talked about are as follows:

1. Traditional business continuity puts in place measures that are largely reactive. There needs to be a move toward a more proactive approach.

2. There is currently no standard approach to business continuity across all organizations.

3. Organizations are not designed with business continuity in mind. Business continuity is often isolated within the organization. There needs to be integration of business continuity across the organization.

4. The constituent parts of the business continuity group within the organization need to be rationalized and rethought to connect with other related functional groups within the organization.

5. There is currently a disconnect between business continuity professionals and those involved in implementing solutions that address continuity problems.

6. The threat of terrorism needs to be addressed more specifically when catering for business continuity.

7. Focus needs to be put on people as the core asset of the organization.

8. Organizations need to be motivated toward better continuity preparation, security, and health and safety.

9. A means of financially justifying these additional or more comprehensive measures must be sought. Insurance organizations need to cooperate with industry to ensure that individuals, economies, and national security are better protected.

Some of these points we have discussed in detail already so I will not labor the points too much further. For the purpose of simplicity, I will label each of these issues as follows:

1. Proactivity;
2. Standardization;
3. Integration;
4. Rationalization;
5. Consistency of response;
6. Terrorism;
7. Human assets;
8. Motivation;
9. Finance and cooperation.

I will outline my approach to each of these issues in tabular form. See Table 2.1. It goes without saying that the implementation of actions defined will form the basis of a new business continuity methodology or framework, that is, continuity assurance.

Some background information and discussion is needed before we move on to define the methodology itself.

2.1 Key Components of Operation

The key components of operation of any business can be grouped as follows:

1. People;
2. Equipment;
3. Workplace;
4. Suppliers;
5. Logistics;
6. Finance.

As I have already stated, people are the key asset of any organization. Without people there is not too much that works. People rely on various forms

Table 2.1
Traditional Methods—Issues and Actions

Issue	Action
Proactivity	Include analysis of prospective risks in terms of identification and analysis of warning signs.
	Responses to events or event warning signs should be scalable and easily retracted if the event does not occur or the risk abates.
	Responses should be enacted based on risk awareness. That is, the response may be executed before the event occurs, as a precaution.
Standardization	Define a comprehensive end-to-end methodology for continuity assurance.
	Define a quantitative rating system for continuity assurance capability.
	Define methods of implementing the continuity assurance methodology (CAM).
Integration	Outline the positioning of the continuity assurance group within a standard organizational structure.
	Define the basis for a corporate continuity assurance awareness and education program.
Rationalization	Define the constituent parts of the continuity assurance group.
	Define the interface points between the continuity assurance group and the rest of the organization, including key or critical interfaces.
Consistency of response	Define the roles and responsibilities for the continuity assurance group.
	Define staff skill sets required within the continuity assurance group and in the wider organization.
Terrorism	Terrorism must be included in the methodology as a key risk to the continuity of any business just as, for example, fire or flooding are today.
	Terrorism is more complicated than standard disaster events and may require more comprehensive analysis of risk and specialist resources.
Human assets	People need to be recognized as the core asset of the organization.
	More robust measures need to be taken to safeguard personal security and safety.
	These measures must be integral to any event response.
Motivation	As a standardized methodology becomes widely used it will become expected by customers and peers that your company meets the standard. You may gain competitive advantage by meeting the requirements of the standard. The expectation may translate into necessity in economic terms.
Finance and cooperation	Once there is a critical mass of organizations following a standardized methodology we may lobby the government and insurance bodies for tax incentives or insurance reductions based on the money spent and the benefit to the wider economic community.

of equipment to do their jobs, whether that equipment is a computer system or heavy machinery or a pen and paper. Clearly, people need a place to work and equipment needs a location to reside, so workplace is important. Without the supply of goods and services products may not be able to be produced or machinery maintained. You must have a means of receiving supplies and distributing product. If this means is physical then logistics are critical. Finally, without money and access to financial markets nothing, including your people, can be paid for. A workforce that is not paid does not stay working for very long.

We have already discussed the fact that most business processes are controlled by IT systems so we will consider the IT infrastructure and systems as the core equipment of the typical organization. This follows from the fact that information is the primary enabler of any business. This information is stored in one of two primary places, people and IT systems.

In terms of safeguarding equipment other than IT systems the measures to be taken are fairly straightforward. In essence these can be dealt with in the same way we deal with computing infrastructure, that is, hardware. Backup equipment, critical spare parts, and physical security will get you a long way. As the IT systems are more complicated to deal with than this we will focus our attention from this point on IT systems as the principle equipment that people use to enable business processes. All other equipment should be covered the same way as computer hardware.

2.2 Events and Event Responses

An event is the development of a risk into an actual occurrence that adversely impacts the continuity of your business.

We have talked about events that affect the continuity of our business and how we respond to these events, but what are they? Well, it is impossible to cover every possible event with an adverse impact to business operation in a generic forum such as this. However, some examples, which are common to every organization, are:

- Fire;
- Flood;
- Localized malicious damage;
- Theft;
- Terrorism and sabotage;
- Explosion;
- Chemical spill;

- Gas leak;
- Disease;
- Earthquake;
- Cyclone or severe storm; to name just a few.

Most of these are fairly obvious but there are some less obvious examples here. Disease for example, something as simple as a flu epidemic can impact your business severely. Some organizations have taken to paying for mass flu inoculations for their staff to attempt to manage this risk.

We can group these events and other events as follows:

- Natural/environmental disaster;
- Explosion;
- Biological agent;
- Hostage situation;
- Threat of action;
- Criminal damage;
- Accidental damage;
- Fault or failure.

So what are the possible event responses that we can employ on the key components of operation to ensure continuance in the case of events such as these? See Table 2.2.

Table 2.2
Key Components of Operation

Operational component	Possible event response
People	Evacuate and, possibly, relocate.
Equipment	Move to resilient system or backup system, replace or recover.
Workplace	Move to recovery site or sanitize primary site.
Suppliers	Invoke emergency/priority supply agreements. Utilize secondary supplier.
Logistics	Use alternate transport mechanism, alternate routes, or secondary supplier. Stockpile and/or relocate stock. Utilize emergency stocks.
Finance	Access emergency fund. Employ backup connections to banks. Move to manual workarounds.

2.3   Facets of Business Continuity

Traditional business continuity is made up of a number of different facets. These are, generically:

- Risk assessment;
- Business impact analysis;
- Strategy development;
- Contingency planning;
- Crisis management;
- Disaster recovery.

This can be represented diagrammatically as follows in Figure 2.1.

2.4   Legacy Approach

At this point I believe it is useful to have a clear understanding of the current typical approach of many organizations to business continuity.

In many cases business continuity is done organically within various parts of the organization without a central business continuity team. Individual

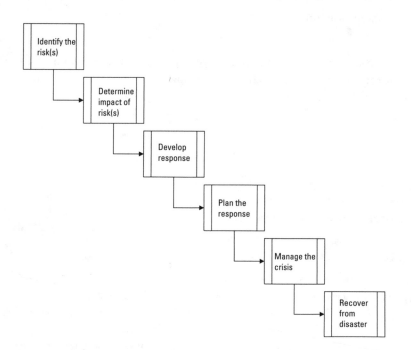

Figure 2.1   Traditional business continuity method.

groups see a need based on experience or learned knowledge and act to address that need by implementing measures that safeguard the organization.

Most business continuity is conducted not as an operational and ongoing function but as a project to assess requirements and implement corrective action. Indeed the analysis and implementation are often done separately, often by different organizations, largely because of the consistency of response problem identified earlier.

The majority of projects are initiated from within the IT organization and revolve around disaster recovery capability. Typical business continuity methods come into play when defining the requirements for creating this capability. Some companies will establish a business continuity function but this function again will typically be located within the IT organization. Furthermore, business continuity functions are generally located within the IT security department. This is done presumably because both business continuity and security are risk management related. Business continuity has not traditionally been located within risk management arms of the business because these teams generally perform financial and accounting-type functions and are concerned with business-as-usual risk management in financial terms. This, however, is changing.

Business continuity teams are generally small and, considering their scope of responsibility are quite isolated, buried under several management layers within the IT organization. Given that true business continuity needs to act over the whole organization this is clearly an inappropriate location in the organizational structure. It means that the business continuity manager has to attempt to exert some authority with business managers who are in a different vertical stream of the organization and are several levels more senior to him or her. Not to mention the usual tension that exists in general between IT and the business, as we have already discussed.

Given the size of the task and the limited resources of a small business continuity team, a project is commissioned that will hire specialists that come in and do the analysis work and rollout business continuity across the organization. Incidentally, the team is usually small because the IT executive has:

- Underestimated the size of the task at hand.
- Is not sure if the business will swallow the real costs associated.
- Has a personal key performance indicator (KPI) linked to doing business continuity or disaster recovery, which is linked to his or her bonus at the end of the year. He or she wants to do whatever the minimum is to meet the KPI, by the required date.
- Has not hired specifically skilled resources but promoted someone from another part of the IT organization, or someone has seen an opportunity to create a business continuity empire within the company and has put himself or herself in the position.

So it soon comes to pass that the team is too small and does not have the required skills to do the task at hand, or is not too sure what the task is exactly. Therefore the problem is passed back up the chain to the IT executive who sets aside some money and commissions a project for the business continuity manager to manage. It has to be done because it needs to be "seen to be done" because of the reasons mentioned above. However, he or she may believe that this could be useful to grow the IT empire, by incorporating the build of a recovery center, or implementing a storage area network (SAN), or perhaps conducting a network renewal exercise. Server consolidation, middleware implementation, high availability, the list goes on and on. All sorts of things can be hidden away in a business continuity budget, mostly because no one knows exactly what business continuity does.

This is all a bit cynical I know and it does not always happen like this, but this sort of situation is commonplace.

Enter the management consultants. Let us take a minute to look at management consultancies to put this into perspective. The major consultancies work in a rather unique way, largely stemming from their partnership structure. Most of these firms operate three core businesses, which we will loosely call, audit, business advisory, and systems integration (IT). Business continuity will typically sit within the business level function of business advisory, which is not generally an IT business area. These guys are usually accountants, MBAs and the like, not IT professionals. You will find that this area is now becoming more IT intensive in most of these firms owing to them spinning off or selling their systems integration arms over the last few years. They have now begun to regrow new systems integration or IT consulting business within the business advisory segment, which they will eventually spin off or sell again. It is a cyclic process and is very profitable, more profitable than the business itself actually. It can be likened to farming in a way. Farmers typically do not make any real money from their farming enterprise but simply break even until such time as their land can be divided or sold at a premium. At this point they get their windfall. The situation with management consultancies is no different. The work they do grows their business and pays operating costs. Then at an appropriate time they sell it and make their windfall.

So a team of business continuity professionals from a leading management consultancy have entered your organization. In many cases they will be accountants with some training in risk management and business continuity. They will look at your business from the top down in the following sequence:

1. Products;
2. Business processes;
3. Systems;
4. People;

5. Data;
6. Physical hardware.

The purpose of this sequence is to establish which systems are critical to the continuance of the business. They will then establish the set of people required to operate the systems and aim to define a strategy for replicating and/or recovering the business processes by recovering the systems and people.

In parallel to this, they will conduct a risk assessment to determine what the likely threats to, and vulnerabilities of, your organization are. This information will also be fed into the strategy debate.

Concerning the six-step process above: To explain this a little further, the business continuity professional will speak to the business and determine what your products are and what their relative importance is to the business in terms of the continuity indicators. Just to remind you, these are:

- Financial;
- Reputation;
- Customer service;
- National security;
- Health and safety;
- Regulatory.

They will then look at what key business processes support the provision of these products. For this they will begin with a basic model for your type of business. For example, the high-level business processes for a telecommunications company looks something like this:

- Billing;
- Provisioning;
- Customer care;
- Supply chain and logistics;
- Sales and marketing;
- Development of new products;
- Operations.

They will drill down on this to lower level processes depending on the exact scope of the project that they have been commissioned to deliver. They will then look at these medium to low-level business processes and determine which IT systems are used to enable these processes, and then who are the key people that are required to operate these systems, both from the business and IT organizations. They will attempt to rate the criticality of these systems determined by the effect on the business process of the system being unavailable. Each consultancy will have its own method of doing this but its generally quite simple. A scale is used with a rating from one to five or A, B, C, or something similar. The highest level may be if the unavailability of the systems stops the business process entirely.

The business continuity professional will then look in turn at the data held in these systems and its effect on the criticality of that system and the individual physical servers on which the system and its data reside. In this way they can assess which servers need to be looked at with regard to resilience or recovery capability.

There will most certainly be situations where the business will accept a limited stoppage of the business process determined by their customers tolerances, which will have been assessed when the products were looked at in stage one.

The problem with all of this largely resides around the particular skills of the business continuity professional. As I have said before these people are unlikely to be IT professionals. This means that when they get into the detail of looking at the individual IT systems and data architecture they almost certainly will find themselves out of their depth. Many of these consultants will barely understand the difference between Windows and UNIX and treat the systems as black boxes that perform a function in supporting the business process, but they do not necessarily understand how. They certainly will not understand how all the systems interface to each other, what the system dependencies are, and so on. This can lead to a less than complete analysis of critical systems. They are also victims of the disconnect we have spoken about between the business and IT organizations. Because they have taken a top-down approach they will have gained most of their information from the business and not IT. The business may be under the impression that a particular system is critical when in fact it is not, because there is a workaround via another system or simply because they think that their job in the company is much more important than it really is. The business continuity professional acting in this role as a business analyst is in many respects bound to report on the information that he or she is given by the company via its staff. There is not enough evidence or knowledge at this level to contradict what the business has said. This conundrum will become the problem of those eventually commissioned to implement the strategy that the management consultancy will define, based on the information collected.

The management consultancy's activities will be to deliver, in general terms, the following:

1. Business impact assessment;
2. Risk and vulnerability assessment;
3. Systems criticality matrix;
4. Continuity/recovery strategy;
5. Framework for system contingency plans.

They are unlikely, however, to have the technical IT knowledge to implement any of the tasks defined within the strategy. Hence they will probably run a tendering process on behalf of their client and select another company to perform this work. Alternatively they may have a go at doing it by subcontracting people or companies that do have the skills to implement. In extreme situations the clients may choose to attempt to implement the strategy themselves. They may use their existing staff or engage individual contractors to work with their staff. Given the high risks associated with projects like disaster recovery the shrewd ones will not attempt this. The risks are high because of the potential impact to the business if you get it wrong. Remember that you are affecting the businesses core infrastructure, and whatever you implement you will have to test.

Enter the systems integrator. This exercise may cost you a lot more than you had planned but these companies are usually well resourced and motivated to deliver. They will more than likely be hampered only by the preceding work done by the management consultants and the internal politics of your organization, not helped in the least by trying to run a project that changes the fundamental landscape of the whole business from within the IT department. They will come across so many technical contradictions to the propositions put forward by the management consultants that you will wonder why you ever developed a strategy.

This is, in essence, the problem with business continuity professionals today. They are great theorists and planners but are often not accustom to the realities of implementation. The methodology set out herein will attempt to address, among other problems, this knowledge gap, by defining the skills required in an integrated continuity assurance team.

It is well to note that while the strategy work has been done top-down, a system integrator will typically go about implementation from the bottom-up on a system-by-system level. We will discuss this when we move onto the CAM. We will see why this is done and also, why it is not always the best idea.

2.5 Continuity Across the Organization

The continuity assurance approach has at its core the principle of integration of business continuity principles and measures across the organization.

I once did some consulting to schools concerning how technology and, in particular, computing should be taught. The best thinking at the time and a concept that I subscribed to was "technology across the curriculum." The principle is simple and is based on the fact that a computer is a tool that is used to work or learn. Because of this fact it was proposed that schools should move away from having computer rooms and computing as a subject in itself. Instead, computers should be available in every class to be utilized in the learning of all subjects. In this way students would learn how to use the tool in a more realistic and comprehensive way. The approach was not generally accepted at the time, because of two factors:

1. To be effective the principle requires that each student have access to a computer at all times. This was not economically possible for most schools.
2. Teachers themselves were not fluent enough with the technology to be able to teach their subjects utilizing latest technology.

This same concept can be applied to business continuity. You could call it "continuity across the organization." The principle is that you cannot successfully do business continuity in isolation. It requires, in most cases, a fundamental shift in the way things are done in every part of the organization. Having a business continuity department that takes over in a crisis situation simply does not work. Nor can the typical business continuity department push through the sort of fundamental changes we are talking about from an isolated position within the organization. What is needed is for the entire organization to be matrixed around a business continuity department placed at a pivotal point in the organizational structure. Unfortunately, the two problems that I encountered with technology across the curriculum are also a problem here. Namely:

1. There is a high cost associated with this approach;
2. Managers will need to be educated in the principles of continuity assurance.

The positive side of course is that it will work, just like technology across the curriculum, which many schools do embrace today quite successfully. It is now generally accepted that isolated computing courses and computer rooms are a less than ideal approach and in many cases do not work. My opinion is that isolated business continuity departments also are less than ideal, and in my experience do not work.

3

Methodology

The following constitutes the CAM.

This is a top-down methodology. Many people believe that the only way something can be successfully introduced, practically speaking, is bottom up, mostly based on the theory that reality only exists at the coalface of an organization. My experience is that this is an approach for people that cannot cope with the complexities of dealing with the big picture from day one. They need to take it on bit-by-bit and hope that at the end of it they have got all the pieces of the puzzle close enough together to make out what the picture is. My experience is that this rarely occurs. They normally end up with all the same problems they had at the start, just with the pieces in a different order. Put it this way, if the big picture is an elephant you may end up looking at an upside-down orangutan.

It should be remembered that all business continuity, including this CAM, is an exercise in risk management. It is not a revenue producing activity, therefore, it is a business overhead. Risk management is nevertheless a unique overhead because it is one that can save you a great deal of money, far more money than any efficiencies gained in the implementation of other business overhead type functions. Risk management is a form of self-insurance and can be justified based on losses that could occur in the case of an event that impacts the continuity of your business and the probability of the occurrence of that event. Budgets should be set accordingly.

This methodology, being an integration of business continuity across the organization, brings with it many other benefits apart from those directly apportioned to minimization of losses in the event of disaster. The implementation of this methodology should have secondary effects in streamlining your organization and increasing business efficiency through the strategic integration of business functions and integrated design of underlying infrastructure.

3.1 Overview

The methodology has two distinct delivery streams, which we will call the offensive stream and the defensive stream. The offensive stream is our proactive measures taken to assure continuity of our business. The seven Rs below represents the core of this stream. Our defensive stream, while still implemented proactively, is principally concerned with the implementation of safeguards for people, property, and information. Our offensive stream acts to ensure that we are prepared for an event and our defensive stream acts to attempt to prevent the event and, if a crisis should strike, ensures that our people are as safe as possible. The defensive stream is essentially comprised of the two Ss: security and safety (see Figure 3.1).

Above these two delivery streams sits a management module. The management module of the methodology addresses the following:

1. Governance;
2. Project management;
3. Communication;
4. Knowledge transfer;
5. Resource management;
6. Strategy;
7. Continuous improvement;
8. Achievement and quality assurance.

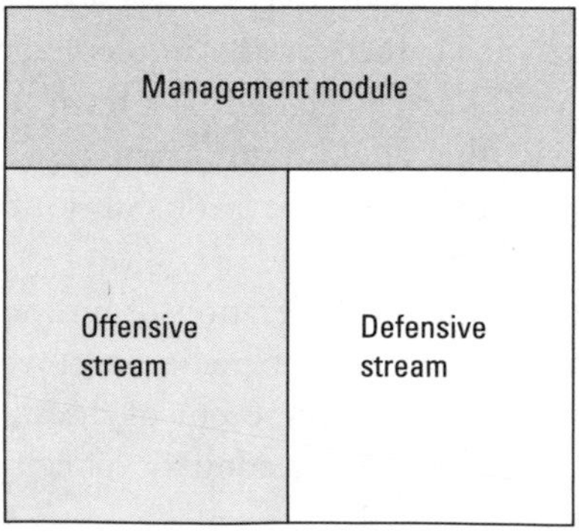

Figure 3.1 Continuity assurance delivery model.

A circular model best represents the complete methodology. I find that cyclic or iterative processes are best represented in this way, as shown in Figure 3.2.

If we start from the core and move outward we begin with governance and strategy. If we use the analogy of the model as being a wheel, the strategy is the direction that we wish to travel in. The methodology, being the complete wheel, is the transport that will carry us there. Governance is the navigation of the journey and ensures that we keep our course and arrive at our destination, on time.

The management hub holds the wheel together. Management controls the day-to-day operation of the continuity assurance machine.

Quality assurance measures progress in terms of achievement and the condition of the machine at any point in time. This function interfaces across and around all other functions contained within the methodology. Quality assurance will also ensure control through the stage boundaries of the maturity model, discussed later.

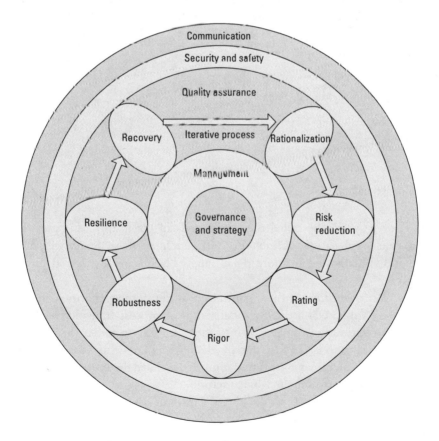

Figure 3.2 The continuity assurance framework.

As the wheel turns we move through the seven maturity levels of continuity assurance. We will talk about each one of these in detail later. The process is iterative because as we move through revolutions of the wheel we are continuously improving or refining our capabilities and reacting to changes in the organization.

The engine is encircled and protected by a ring of security and safety. These are the defensive elements of the CAM.

Lastly, encircling all elements of the methodology is communication and knowledge transfer. Consistency and clarity of message and education of all people involved in the organization is imperative for the methodology to be used successfully.

3.2 Offensive Stream

Methodologies should be simple. The core of this methodology is what I will call the seven Rs. The seven Rs are:

1. Rationalization;

2. Risk reduction;

3. Rating;

4. Rigor;

5. Robustness;

6. Resilience;

7. Recovery.

They may also be called the seven levels of continuity assurance. I will give a brief explanation of each in turn. Each of these levels will be discussed individually and in more detail later. These can be referred to as R1 through R7.

R1: Rationalization of the organization to harmonize security, continuity, and recovery functional areas. Rationalization of process to avoid overlap and ensure that continuity is integrated into the organization rather than added on to it. Rationalization is the first step on the path to continuity assurance. Your organizational structure is your foundation. If you have the underlying structure wrong you will undermine the whole method.

R2: Risk reduction is both the identification of risks to the business and the measures the organization determines to be put in place to reduce each risk identified. This is the second step, it can neither occur before rationalization or after rating. Without the organization ordered it will be impossible to see the wood for the trees. It is also necessary to reduce or eliminate risks in order to accurately rate true criticality of processes, people, and systems.

R3: Rating, of people, process, and system to ensure that the organization is aware of its critical components and assets for conducting business. This should also expose any weaknesses. This process needs to occur after rationalization and rating and before rigor. If you have not thought about your structure you are not ready to look at its individual components in more detail. You may not even know what the components are. Again, if you have not first attempted to reduce risk you are not getting a true picture of the criticality of the components. Without clear criticality ratings you will be unable to determine where processes need to be strengthened at the rigor level.

R4: Rigor of process. Without the processes in place to manage the configuration of your components and to control your business processes you do not have a clearly defined position to recover to. Rating needs to have occurred to highlight areas that should be reinforced process wise. Without clear and effective processes it is impossible to know which supporting systems will need to be reinforced or made more robust.

R5: Robustness of architecture. Given the information derived from previous levels it should be possible to determine the vulnerabilities in your infrastructure and take an integrated architectural approach to correction. Note that there may be technical prerequisites for entering or exiting this level. This work can clearly not occur prior to having corrected gaps, where possible, by strengthened processes. You will also need the increased control gained over your environment by these processes in order to safely manage any fundamental changes to the architecture. Before looking at resilience we must first make sure that our underlying architecture is sound so as we are not replicating something that is less than ideal.

R6: Resilience of systems. Now that we have achieved all that we can through strengthening process and underlying systems architecture we may introduce a level of insular resilience to our critical components. Note that resilience can apply to more than just IT systems. Remember that both people and systems carry information. It is this information that we need continuity of access to. A level of geographic diversity to any resilience design may mean that you avoid recovery if the event is localized. Good resilience will aid in your ability to be proactive to an event that you feel may occur but has not yet occurred.

R7: Recovery of business processes. Recovery is what we are left with when all other preventative measures fail to work because the event is too widespread or severe. Any recovery operation has an associated risk and if you are at this point you will almost certainly be experiencing an interruption to your business. If you have achieved success at all in the previous levels of continuity assurance you will only need to recover in the most extenuating circumstances.

3.3   Defensive Stream

The defensive stream of the CAM is characterized by the two Ss.
    The two Ss are:

1. Security;
2. Safety.

I will give a brief explanation of each in turn. They may be referred to as S1 and S2.
    S1: Security of people, property, and information. If your organization is secure and you have taken proactive measures to prevent or deter an event you will have lessened the risk to your business. People need to be safe and feel safe in order to be able to work efficiently and be focused on the continuity of your business. Information is the core asset of any business and must be protected. Information is stored in property, such as computer systems, and in people.
    S2: Safety of staff and public. Your staff is your most important asset and must be looked after. By taking care of your staff you may avert an event that impacts your continuity. Accidents do happen but we can take measures to make their likelihood less probable. If staff are unavailable due to injury or sickness you do not have access to the information that they carry. This may cause disruption or in the worst cases interruption to your business operation.
    S1 and S2 are integrated across R1 to R7 in that there are elements of security and safety at every level of the maturity model. We will discuss these elements in some detail when we define the characteristics of each maturity level.

3.4   The Maturity Model

The seven Rs of continuity assurance are intended to be utilized as a maturity model. As indicated by the brief descriptions above, the levels should be tackled each in turn before proceeding to the next. Later we will define the entry criteria and achievement required for progression to the next level. The methodology is designed in this way to leverage maximum efficiency from the actions undertaken at each level.
    Below you will see the continuity assurance pyramid, a standard hierarchical representation of the maturity model. An organization will progress from the foundation level to the pinnacle, creating business continuity capabilities along the way. Once the pinnacle is reached a phase of continuous improvement is entered in which the organization revisits each level with the benefit of having been all the way through the levels once. At this point they should have a solid grasp of the big picture and may tweak their capabilities at each level to reflect their now holistic approach.

Once the structure has been fine tuned in holistic terms it is time to begin again from the foundation level reassessing capability and where necessary employing new capabilities to reflect the ever-changing business environment. See Figure 3.3.

If the seven Rs comprise the offensive characteristics of continuity assurance and are represented by a pyramid then the two Ss, as the defensive components of the methodology, may be represented by casing stones present at all levels. The casing stones line the outside walls protecting the underlying structure.

S1 and S2 are not levels of the maturity model per say. They should exist at every level. Each level has specific requirements around security and safety. Without meeting these requirements it may not be possible to move to the next level.

As we move up the levels we are continually creating more capabilities in continuity assurance. Each level builds upon the foundation laid by the previous level and thus the capabilities build upon each other. Each capability is ordered and is part of the overall strategy of the model. A capability at level 4 may not logically exist without capabilities at levels 1, 2, and 3 first being attained. You may be able to create a capability at level 7 from the onset but this capability will not be strategic in nature and will not be of the level of quality it would have been if you had progressed through the sequence of the preceding levels of the model.

3.5 Achievement and Quality Assurance

Quality assurance is about four key factors:

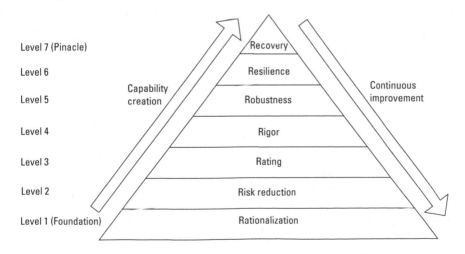

Level 7 (Pinacle) — Recovery

Level 6 — Resilience

Capability creation — Continuous improvement

Level 5 — Robustness

Level 4 — Rigor

Level 3 — Rating

Level 2 — Risk reduction

Level 1 (Foundation) — Rationalization

Figure 3.3 The continuity assurance maturity model.

1. Achievement;

2. Control;

3. Durability;

4. Efficiency.

Each level of the maturity model has defined within it several KPIs. These KPIs are designed to be assessed quantitatively. Each KPI has a minimum score that is necessary for the achievement that goal within any given level. This does not mean that scores cannot be improved on in subsequent iterations through the model. Continuous improvement should be sought and is integral to the functionality of the CAM.

The overall rating of KPIs across all levels of the model forms what we will call the continuity assurance achievement rating or CAAR. This rating is intended to be a measure of overall business continuity capability. The capability ratings for each level will form the basis of the exit criteria from each level in turn and the entry criteria for the following level. In this way we ensure a measure of quality control at the stage boundaries between the seven maturity levels of the model.

Quality must be applied to all elements of the CAM. Quality is about ensuring that a strategic approach is taken to all work with tactical activities only being performed to fill a gap until the strategic solution becomes available. This means that strategic and tactical solutions or plans are always executed in parallel. This focus on strategic goals I call durability. Tactical solutions by definition have a short lifespan. Strategic solutions are intended to be more durable. They are better thought out and are designed to react to changes in their boundaries. Yes, it often occurs that, for example, a piece of software is designed and implemented to fill an immediate need and is intended to be replaced; however, you may find this piece of software still operating after many years. This may be true but there is usually a cost associated with the short-term nature of the thinking behind the solution. You may find that the software has had to be added to and modified constantly throughout its life span, leading to an exceedingly high cost of ownership. You may also find that the solution is not exactly fit for purpose and means that there is additional activity taking place somewhere else in the organization to compensate for this shortcoming. Essentially, because the solution will not be fully thought out and integrated into the operation of the organization it will be inherently more inefficient over time. You may feel locked into such a solution if you leave it in place for too long. You may find yourself trying to expand on it to make it fulfill a more strategic role, but it is likely that this sort of thinking will lead you down a dangerous path. You cannot build a skyscraper on the foundations of a three-bedroom bungalow. It is sometimes better to start again, or better still do the work in parallel. Trying to make a tactical solution into a strategic answer is inherently inefficient.

Quality in terms of the use of a methodology is also about adherence to that methodology. It is useful to conduct regular quality reviews across the organization, continuity assurance department, or continuity assurance project to ensure that the methodology is being properly followed. It is best to engage the services of a third party for these types of quality audits. It is difficult for people to realistically assess what they see every day. In any event, a new opinion is always worth having. Listen to criticism and make allowances for what you think the auditor does not understand about your situation. Try to be self-critical of your work and take notice of how others perceive the quality of it.

Testing is something that is often grouped as part of quality assurance and in textbook terms it is. However, I view this as more of an operational task with regard to business-continuity-related activities so we will talk about testing when we discuss the individual requirements within the relevant levels of the model. The role of quality assurance in testing, within this methodology, is to give oversight and ensure the integrity of the tests conducted and results obtained.

Quality principles should be applied to all facets of the model with particular attention to any deliverables, be they documents, software, hardware, tests, training, procedures, or others.

It is important to mention that the CAM advocates quality assurance as opposed to quality control. Quality assurance measures quality at all stages of the development of the deliverable whereas quality control typically will assess only the final deliverable. This means that if there are errors found in the final deliverable there is rework needed to the degree that the deliverable may need to go back through the entire development process in order to correct the problem. If we assess quality across the development life cycle for all individual components of the deliverable then errors are found earlier and reworked along the way. This approach is generally considered more efficient and will ultimately result in a higher quality product. This is essentially the principle of total quality management (TQM).

We will discuss quality assurance in more detail later, particularly with respect to project-based quality assurance.

3.6 Knowledge Transfer and Resource Management

Information is the core asset off any business and the principle repositories of your company's information are people. It is important that the management structure is organized in such a way as to optimize the access to and utilization of this information for the betterment of your business.

To this end I recommend that a less conventional management structure be utilized for continuity assurance. You may wish, based on the information given here, to extend the use of this resource management framework to other parts of your organization.

The management framework outlined below, in addition to its general management advantages, addresses many of the requirements of the methodology in terms of knowledge transfer. Most management structures are quite inefficient in their ability to transfer knowledge to their subordinate members, the wider organization, or in the case of a contracted firm, the client and the client's employees.

Many feel that the best way to achieve knowledge transfer between, for example, a project team and business-as-usual staff is by creating a matrixed organization made up of a combination of resources from both camps. Roles are generally duplicated with all key roles having a project responsibility and a business-as-usual responsibility. It is essentially the system of shadow roles. Others favor the handover approach. I do not subscribe to either of these views; mostly because I have never seen either of them work.

The typical scenarios in which knowledge transfer becomes critical are:

1. Transition from internal project to business-as-usual;
2. Transition from external agency to internal operations;
3. Transition from departing staff to joining staff;
4. Transfer between primary responsible and secondary responsible.

The key problem in scenario one above, and to some extent in scenario two also, is that project people are generally of a different mindset than business-as-usual people. If you take a business-as-usual person and put him or her in the middle of a project he or she is more than likely to have difficulty coping with the pace of work, delivery focus, and increased responsibility that comes with any project. Conversely, if you take a project resource person and ask him or her to do something with perhaps a wider business impact than an individual project person he or she often will not cope well because too much is going on that is outside of his or her direct control.

One of the key problems that I have mentioned previously is that of appropriate staff skill levels. This framework also attempts to address this problem.

Before I move on I would like to share a saying that I have regarding resourcing. I have found this to be more than accurate. It basically says:

- Qualifications are good.
- Experience is better.
- Intelligence is best.

Both qualifications and experience can be gained but intelligence is innate. A person with superior intelligence can often make more sensible decisions than someone with all the qualifications and experience in the world. Remember this when you are resourcing your teams and it will serve you well. Do not discount people based on lack of qualifications or experience. Some intelligence, aptitude, and enthusiasm will often be a better bet. Of course an intelligent person with experience and qualifications is best of all but having one or two of these does not necessarily imply the others.

A key problem with scenario two is that where you have a shadow team in a client/consultant situation the consultant will always end up reporting to his or her shadow on the client side. This is counterproductive given that the consultant is more knowledgeable than the client, otherwise he or she would not be there at all.

Scenario one would generally be done via a shadow team, probably in the later stages of the project but in some cases from start to finish. Scenario two would be done using a phased approach to handover, or a shadow team again. Scenario three would be via handover, that is, staff overlap or a document-based handover, or, quite often not at all. Scenario four would be via staged training or some level of handover, generally speaking.

Table 3.1 details the continuity assurance approach to these scenarios. A detailed explanation follows the table:

3.6.1 Team Blending

This method is based on employing a different structure to a team than what we are typically used to. I call this the hospital model as it is based on the way doctors are organized in a hospital, based on many Western medical systems (see Figure 3.4).

Table 3.1
Knowledge Transfer Scenario Approaches

Scenario	Approach
1	A blended project team whose management remain consistent throughout the transition to business-as-usual (see below).
2	An apprenticeship-based method (see below).
3	A shadow approach with the incoming resource shadowing the outgoing resource for a reasonable period; or, where this is not possible, a handover to the secondary and a shadowing of the secondary for a reasonable period.
4	Parallel training of these resources from the beginning leading to good up-to-date knowledge of the role. A handover will still be required to the secondary. Once handed over, the primary should shadow the secondary for a reasonable period. Every key role should have a primary and secondary resource; this is how we affect resilience of the information stored in the people of your organization.

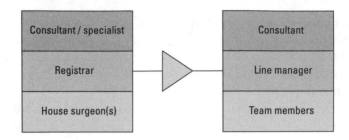

Figure 3.4 The hospital model.

There are typically three levels or grades of doctor in a hospital. They are separated into teams, which specialize in a particular area. The consultant leads the team and is usually a specialist in his or her field. The consultant is not generally an employee of the hospital and is not always on site at the hospital or working on hospital-related work. Next is the registrar who takes the role as manager of the team on the ground. The registrar deals with the day-to-day operation of the team which of course involves doing rounds, operations, and other duties just as any other member of the team. Below the registrar are one or more house surgeons who do the routine work required and escalate to the registrar where appropriate. The registrar may also escalate to the consultant when required. Both the registrar and the house surgeons are normally employees of the hospital. Below the house surgeons are any trainees, that is, interns or the like.

The normal consulting model in business is the reverse of this, where the consultant is brought in underneath the relevant manager within the organization. This model does not work well because you are always in the situation where the consultant knows more than the person to whom he or she is reporting. The consultant will also be constrained in what he or she can do or suggest by the whims of the manager. Yes, the manager can learn from the consultant but why does this have to occur this way around. The manager can learn just as well working for the consultant. In fact this is the more natural approach. Yes, the consultant is not as familiar with the organization as the manager is. But this is the case anyway. So we would normally have a flow of information about the organizational specifics from the manager down to the consultant, so why cannot this also work the other way around? The manager can just as easily deliver this information upwards to the consultant who then has all the information required, and more diverse experience, upon which to base a decision.

If we employ this approach, with the consultants on top, then we avoid many of the transitional problems encountered in handing over projects into a business-as-usual organization. We essentially enable the business-as-usual structure to deliver projects by inserting consultants over the top of the existing structure and driving the specialist information needed downward through the structure. The consultant(s) may be part of this organization full-time throughout the life of the project; but may generally only be around from time to time during other periods. In this way the organization remains fundamentally the same from project to business-as-usual.

Of course you may need to pad out your teams in order to cope with workload but your overall headcount at any one time should be less than if you remained conducting your business with separate teams for projects and business-as-usual. You may need contractors from time to time to perform specific tasks where you do not need the skill resident in house. Do not confuse these contractors with consultants, the two are quite distinct. A Java developer is a contractor and a business continuity advisor is a consultant, generally speaking. Though if you were running a Java development team the Java developer may well be a consultant.

The consultant may work more than one organization at a time but this is after all the nature of consulting. The benefit you are getting is one of experience, and not experience of only one organization or project but of many projects across many different organizations. You cannot hope to keep this sort of experience in house and to do so would be to immediately reduce the knowledge of the consultant anyway, over time, because he or she would no longer be working on multiple projects in multiple organizations.

To reiterate, the concept is to retain a regular time slice from the consultant. The consultant is in charge of the department but the full-time manager beneath him or her deals with the day-to-day management of the team. The consultant makes the critical decisions and takes part in the critical discussions. You need to empower the consultant to help you with your business. If your managers had all that was required then you would not hire consultants in the first place. If your managers had that level of experience then they would be consultants, and so on and so forth.

If you are worried about continuity of resource, do not be. This is why you have a governance model. People involved in governance operate in partnership with the consultant and do not need to know more than the consultant. They simply represent various views. These will typically be senior people within the organization with views on overall strategy and the like. Those in the governance structure will typically be permanent employees of the company.

The only type of organization where this model will not work is within the delivery arm of a consulting firm itself, for obvious reasons.

3.6.2 The Apprenticeship Method

Aside from the governance of the project or activity the client resource works as an apprentice to the consultant resource. The consultant can give the client a range of work across his or her area of responsibility and get him or her involved at the ground level. Only in this way does the client get true exposure and a detailed understanding of the work involved in managing the team.

There is some overlap here with the blended team in that the line manager acts as an apprentice to the consultant.

The apprenticeship method as described will only work where the project is run entirely by an external organization. If this is not the case then the blended team approach should be used.

3.7 Communication

With business continuity initiatives in particular, consistent and complete communication across the organization is imperative to success.

Communications in this context is akin to a public relations and marketing exercise. There is no initiative that takes place within an organization, with the exception perhaps of a merger, which involves as much change to the organization as does business continuity. For this reason good communications is a critical success factor of any such project.

To effect so much change across an organization you must have the buy-in and cooperation of a significant number of people, key and otherwise, throughout the organization. Because of the nature of continuity assurance everyone in the organization becomes part of the solution. You cannot be part of a solution to a problem you do not know about, so part of communication is awareness in this sense and this will also go hand in hand with knowledge transfer.

There are many ways to effect communication. The most common methods are:

- Intranet sites;
- Company or department newsletters;
- E-mail;
- Group briefing sessions;
- Workshops and seminars;
- Individual briefings.

These are all pretty straight forward. Intranet sites are good but can give the wrong message if they are not updated. Briefing sessions work well but you need to get the audience right.

Be warned. Communication is not an activity that can be done piecemeal or half-heartedly. It is a demanding task and will require dedicated full-time resource and materials. If you are issuing newsletters or doing an intranet site get a professional graphic designer to do the work. Your best resource for communications will come from a marketing background because communications is a marketing role. Pay attention to this area and get it right, if you make mistakes here the consequences can be severe. It does not matter how well you are doing with achieving your technical capabilities. Perception is everything.

The CAM requires that you have a communications road map. This road map should be in step with your project plan to the extent that people who need to be involved know about the project before you ask for their involvement. Think about whom you need to speak to and when you need to speak to them.

There is a common rule of advertising which says that your customer has to be exposed to your advertising at least three times before you have any chance of converting to a sale. The same is true here. Your target must see or hear about your initiative multiple times before you approach them. Ensure this and you will have an easier job getting the cooperation that you need. Your roadmap should set out key functional areas and key people where possible. It should lay out the sequence in which you will need to speak to them. You will not implement continuity assurance capabilities without the help of many people throughout the organization. Best that they know what you are looking for before you ask.

Of course once the capabilities are in place everyone will need to know what they have to do if there is an event. Making sure this is known and raising general awareness of continuity within the organization is also part of this function. It is well to note that rehearsal and testing activities are some of the most effective means of communication where business continuity is concerned.

3.8 Problem Solving and Root Cause

A methodology by itself is not going to answer all your questions or provide solutions to all the problems that you will encounter in trying to increase your business continuity capability. The CAM will provide a framework with which to work within and give guidance on what to do and how to do it. Every organization is different and you will be faced with your own specific challenges. For this reason I want to talk a little about problem solving and root cause analysis.

Problem solving is a skill and it is one that unfortunately most people are not very good at. One approach to problem solving is root cause analysis. Root cause analysis is a requirement of the CAM.

Briefly, this involves analyzing any given problem to determine the underlying cause. If we treat the cause we solve the problem. Problem solving in this way will hopefully minimize recurrence and also minimize the number of additional problems you introduce while tying to solve the initial problem. You may even solve a few other problems that you did not realize you had.

Most people do not use this method of problem solving. It requires some intelligence and time to perfect. For example, what is the root cause of the pension crisis in the United Kingdom at present? Is it falling stock markets, people not having paid their contributions, or company executives running off with pension funds? No, none of the above. The root cause is related to the aging population and the percentage of workers verses the percentage of pensioners. Is the root cause that not enough people are having children, and have not been for some time? Why are they not having children? Is the root cause that mortgage finance is geared on double incomes now instead of single incomes? Well I do not know the answer to this one but you get the idea. You see that root cause analysis can be quite complicated. Luckily, we don't need to analyze and solve the problems of the world here. We just need to address the continuity problems in our organizations and solve the problems that we encounter in implementing capabilities to safeguard this continuity of operation.

4

Level 1: Rationalization

This foundation level of the continuity assurance maturity model focuses on the rationalization of your organization.

The following segments of the organization need to be blended together and revisited for overall scope overlap. The segments are:

- Physical security;
- IT security;
- Business continuity;
- Disaster recovery;
- Health and safety;
- Risk management.

The resulting group will be the continuity assurance department or team. The position of this group in the overall organizational structure is of utmost importance.

In many organizations, business continuity and disaster recovery are situated within IT security. This may seem appropriate as all of these functions have a basis in risk management and all will have a heavy focus on IT; however, continuity assurance needs a level of authority over the entire organization that cannot be affected from this isolated position within the organization. It should be noted that some companies have begun to move in a similar direction with this and are increasingly viewing BCM as a wider business issue.

Few corporate structures are approached in any type of strategic way. Most evolve over time and are driven by the relative power of key individuals within the organization and the politics of how these individuals interact with each other. For example, if an individual is at a particular management level and is in charge of a particular division that is being disbanded because of some shift in the company's strategy, that individual will be given a new role. You cannot put that person in a situation where he or she is reporting to someone who is his or her peer so whatever role you assign, the person will remain at the same level in the management structure. The problem is that the role you assign will not necessarily be logically at that level within the organization. So you have a problem. You may end up with something like sales and marketing, product development, operations, quality assurance, service delivery, and market analysis or business improvement or some such concocted term. These functions clearly belong at a lower level beneath the relevant sections of the organization. This problem will develop at every management layer within the organization and eventually result in a segmented, unordered, and illogical structure. The nature of this resulting structure will lead to inefficiencies in the operation of the business and potential risks that can reveal themselves as gaps in continuity. This situation must be addressed at this level of the continuity assurance methodology.

Implementation of the measures detailed at this level of the maturity model will increase your organization's inherent resilience in a crisis situation and create a foundation, which you can build continuity assurance capabilities upon. You will also find that the implementation of these measures gives rise to many secondary benefits in the form of efficiencies in your organization.

It should be noted that the situation above, where we have managers reassigned roles at the same level as a previous role, would not occur where the blended team is in effect. Positioning a consultant as the team leader means that this person, politically, does not need to be reassigned. His or her contract may be terminated with little problem. In addition a consultant in a particular field cannot be reassigned another unrelated role as the model says that the consultant must be a specialist in the field that he or she is leading. Clearly a new field demands a new consultant. This approach clearly gives additional control over the evolution of the organizational structure.

4.1 Organizational Positioning

Figure 4.1 represents the proposed positioning of the continuity assurance team in a typical organizational structure. Note that this structure should remain true for organizations of any size. The delivery and operations functions represented by the lower-level groups in this structure are indicative only, and are only listed for the completeness of the example.

Given that the continuity assurance function has oversight responsibility for continuity across the entire organization the team needs to be similarly

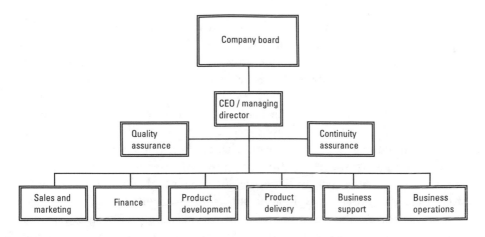

Figure 4.1 Continuity assurance positioning.

positioned over the organization to be able to fulfill these responsibilities. As we have discussed previously, the continuity assurance team needs to be in a position of authority over all business functions. The continuity assurance manager or consultant needs to be a peer with the heads of all major business functions within the organization.

You will note that quality assurance is placed at the same level as continuity assurance in the example structure above. Quality assurance is a similar type of function to continuity assurance in that both functions have an assurance responsibility over the entire organization. Quality assurance also has a role to play across the continuity assurance function. These two teams must jointly affect quality and continuity across the organization. They are both logically separated in the structure from what we would typically call the delivery and operation streams of the business. They are both management functions that the CEO/managing director may act through. They may also report their findings and opinions independently, directly to the CEO/managing director. While the CEO/managing director has as his or her principle goal the commercial success of the business, these functions keep him or her in check to ensure that quality or continuity is not being sacrificed in the pursuit of this goal.

It should be noted that there is one function missing from the above structure that should perhaps also exist at the same level as quality assurance and continuity assurance. This function is the project management office (PMO). Too many times we see projects commissioned by various parts of the organization that overlap or seem to be done out of any logical sequence. The problems caused by such sequencing issues can have drastic effects on continuity. Positioning the PMO at this level within the organization can correct these problems. There needs to be high-level coordination of projects such that these projects are executed in a logical order, information is shared between the projects, tasks are coordinated to minimize risk to the business, and cross-impact assessment is conducted between the projects. When I talk about cross-impact assessment I am talking about changes effected to the organizations infrastructure or landscape by one project that may have an adverse effect on, or impact the delivery of, another project. An example may be a server consolidation project running in parallel with a disaster recovery site build. Clearly if the production infrastructure architecture is being changed it is not a good time to try to establish a recovery site. Apart from the obvious consistent rework required, running two activities such as this at the one time is high risk to the continuity of the business in itself. This situation shows that by trying to implement business continuity capabilities without coordination across the organization you could end up causing a disaster instead of preventing or protecting against one. There are clearly some activities that should not be done in parallel. It is the role of the PMO working in association with the continuity assurance team to ensure that projects are executed in a sequence that does not itself increase risk to the business.

4.2 Team Structure

The structure shown in Figure 4.2 represents a logical order of functions within the continuity assurance organization. You will see that the continuity assurance consultant leads the team, supported in the day-to-day management of the team by the continuity assurance manager.

In line with the blended team approach, the continuity assurance consultant will be a specialist in all facets of business continuity, including disaster recovery. This person will have grounding both in business risk management and IT. It is possible that this person may not be a full-time resource.

You will see that the continuity assurance consultant has a dotted line to the quality assurance manager and the project office manager for reasons previously discussed. These resources are peers to the continuity assurance manager. There will be a high level of interaction across these three teams. The PMO has a general risk management role to play and is a primary information source for the continuity assurance team. The PMO will also clearly play a part in the management of projects commissioned within the continuity assurance team. A level of quality assurance is required by the CAM and will be affected through this team acting across the continuity assurance organization.

Note that while this structure, as shown in Figure 4.3, is hierarchical or vertical, any projects should be run horizontally across the organization matrixing resources from each team. The project manager should not be a member of any of the teams and will typically be recruited specifically for the management

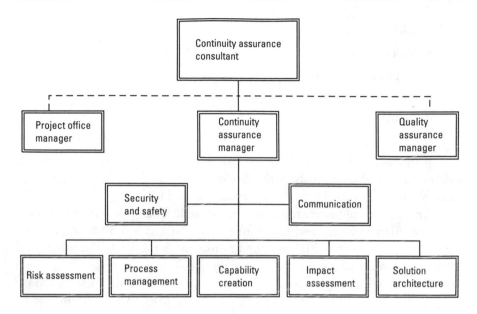

Figure 4.2 The continuity assurance organization.

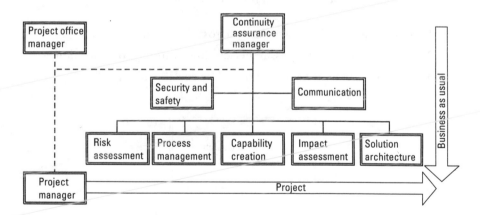

Figure 4.3 Matrix management.

of the project at hand. In some cases the continuity assurance consultant may take on the role of project manager for such initiatives.

The continuity assurance team will consist of the following high-level functional areas:

- Security and safety;
- Communication;

- Risk assessment;
- Process management;
- Capability creation;
- Impact assessment;
- Solution architecture.

It should be remembered that some of these functional areas represent reasonably large teams, especially in the case of security and safety. The lead positions in these teams are, in most cases, executive level roles.

The security and safety function is responsible for all physical and IT security in addition to health and safety functions. Depending on the size of your organization this position may warrant a consultant in this field. The lower-level security and safety group may look something like this (see Figure 4.4). Note that these functions replace the groups currently performing these functions within your organization. Current groups should be harmonized with this structure. There should be no duplication of these groups within any part of your organization.

The communications function is responsible for:

1. Continuity assurance awareness across the organization;
2. Fostering a continuity assurance culture within the organization;
3. Distribution of information regarding specific initiatives;

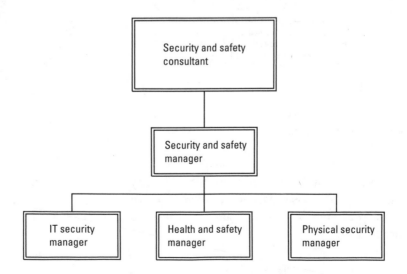

Figure 4.4 The security and safety organization.

4. Marketing of initiatives throughout the organization;

5. Ensuring that all parties are aware of their roles and responsibilities in a crisis situation;

6. Ensuring that training is provided to staff across the organization where required;

7. Coordination of knowledge transfer activities.

Note that knowledge transfer is inherent in implementing this type of continuity assurance team structure. This team merely retains coordination responsibilities for this function. Knowledge transfer is integral to the methodology in general. All functions have a general responsibility to ensure that knowledge is imparted to those that require it.

The first task of the communication team is to communicate the changes to the organization made in implementing the CAM. This will begin with the development of a communications road map to determine the priority order of people to interface with across the organization.

The risk assessment function is responsible for analysis of all business processes, people, systems, and locations in terms of identification of risks and vulnerabilities. In many cases this team will need to work closely with the security and safety function where risks are security or safety orientated. A risk may be as simple as coffee cups left in the data center or as complicated as shortcomings in coordination of projects in the PMO. This is a broad function and there is a lot of ground for this team to cover. Risks and vulnerabilities should be continuously sought out and monitored. In addition to identifying risks and vulnerabilities this function will also determine the projected probability of occurrence of events that exploit these risks and vulnerabilities. There may also be a level of testing required here to identify vulnerabilities. This testing may take the form of intrusive probing of, say, physical or logical security.

The impact assessment function follows the risk assessment and is concerned with determining the impact to the organization of events occurring that exploit these risks and vulnerabilities. This function will look at the organizations products, business processes, and so forth using risk and impact information to rate each component of the organization's infrastructure in terms of criticality. Remember that the stoppage of the smallest insignificant process can have a ripple effect. For example, interruption to the supply of paper could cause your entire billing process to come to a halt. If we can identify the risks and understand the impact of events then we are in a position to identify how to best address the risk, how we defend against it, and how we deal with the event if it occurs.

Solution architecture is responsible for the design of solutions to address the needs identified by the risk assessment and impact assessment teams. These solutions may range from alterations to the IT infrastructure to changing building layouts for improved safety or security. This group will require a very diverse skill set and may utilize a large number of contractors to undertake specific work. This group will be heavily involved in working with all the functional leaders to define strategy based on what is practically achievable. As both risk assessment and impact assessment are ongoing iterative processes so to is defining solutions for the problems identified. This is not a function that can be performed solely as part of a strategy or implementation project. The CAM requires continuous improvement hence this function must be active at all times.

The capability creation function is concerned with the implementation of solutions defined by the solution architecture group. Again, these should in principle not be run only as projects but as ongoing resolution of potential problems identified to the continuity of the organization. This function will need to work closely with all elements of the business and support functions to ensure that solutions are implemented in an integrated way across the organization. Just as in a software development project where defects are identified and resolved so to must risks be identified and addressed as part of the business-as-usual functions within your organization. The resolution of a defect is not a project and neither is the treatment of a risk. It is merely a task that must be performed.

The process management function looks after the optimization and enhancement of processes within the organization. These include but are not limited to continuity plans, disaster recovery plans, crisis management plans, and the like. The process of identification of risks, impact assessment, solution design, and capability creation are also defined within this function. Whereas capability creation is concerned with physical changes to the landscape of the organization; process management is concerned with definition and implementation of the procedural side of the equation. This team must also work closely with general business and support functions within the organization to ensure that processes are sensibly defined, well implemented, and regularly tested. These overall processes are the glue that will hold any physical solution in place. Again, processes must be continually reassessed and optimized in line with the objective of continuous improvement. In many cases plans will need to be developed by the business or the support functions because only they will have the detailed knowledge to be able to do this. In these cases the process management team will oversee the creation of the process and ensure that it meets the objectives set out. The team will also ensure that the process is continuously tested and updated in line with any changes to the underlying landscape of the organization.

4.3 Roles and Responsibilities

Table 4.1 sets out the key responsibilities of each functional area. Also shown is the mapping of functional area to responsibilities under each area of the CAM.

These are represented by R1 through R7, S1 and S2, and M representing general management functions under the methodology. Note that all functions have secondary responsibilities across all segments of the methodology. These are not listed here as these overarching responsibilities should go without saying.

4.4 Resourcing

It is imperative in this model, and in any organizational model for that matter, that we have the right resource in the right role. Any process is only as strong as its weakest link and an unsuitable person in a role that he or she is not capable of can be the weak link that causes the whole business process to break down.

As part of the process improvement requirements at this level of the model, organizations are required to look at only one key process area, which is human resources.

We will focus on recruitment, as this is the core process that this support area performs. Once we have a person on board and find that he or she is not suitable, the course of action that we take is often determined by what is allowable under the relevant employment laws for your country, state, or area. I do not think it is possible to discuss these in general terms so I will not attempt to. In any event poor recruitment is generally the root cause of any such staffing problem.

Employers should first be aware that we now live in a global community and the best resources will demand international rates. International rates basically fall to whatever the work would be paid in the United Kingdom or United States. You can get away with a degree of factoring for local cost of living but given that you will often be in the situation where the person is relocating or living away from their principle home this cannot really be used sensibly. At the end of the day one simple rule applies. You get what you pay for. If you want to pay half the international market rate then you will get someone that is half the resource and you will probably have to hire two of them to get the job done. This plus the management overhead of two or more people means that this is just not cost effective. Things will cost what they cost, if you cannot afford the resource then you cannot afford the project.

The next problem in human resources concerns the use of recruitment consultants. Be sure that the consultant you are using has some knowledge of the subject area that you are recruiting for. Without a reasonable level of understanding the consultant will be unable to sort the wheat from the chaff and the honest candidates from the exaggerators. You may find that the best candidates do not even reach you because of poorly worded advertisements or their application being halted due to the consultant not understanding the content of their resume. If your organization is small, utilize niche agencies for specific roles as

Table 4.1

Functions and Responsibilities

Functional Area	Primary	Description (not exhaustive)
Security and safety	S1, S2	All elements of: Health and safety; Physical security; IT security.
Communication	M	Information distribution; Information dissemination; Knowledge transfer coordination; Continuity assurance marketing; Elements of management reporting; Cultural change; Awareness.
Risk assessment	R2	Risk assessment; Vulnerability assessment; Event probability; Testing for assessment.
Impact assessment	R3	Impact assessment; Component analysis; Configuration management; Criticality rating.
Solution architecture	R5, R6, R7	Strategy determination; Solution definition/design; Problem management; Design assurance.
Capability creation	R5, R6, R7	Solution implementation; Solution testing; Problem resolution.
Process management	R4	Continuity assurance workflow; Contingency plans; Crisis management plans; Disaster recovery plans; Incident management; Test management; Change management; General process improvement.

these companies are likely to have consultants with specific experience resourcing for the type of role you are attempting to fill. Avoid entering into standing contracts as your next role may require a different type of agent. Have the department requesting the role deal directly with the agent. If your organization is big enough it may be cost effective to bring this function in house. Hire specialist people for each general area of recruitment (e.g., IT, finance, etc.). These will need to be senior people who have been IT or finance professionals in the past. You need someone with the breadth of experience to understand all types and levels of the role that you may need. This recruitment group should sit centrally in the organization, within HR, and act as an internal recruitment agency. They can also assess staff for transfer into other positions within the company and deal with other internal applications. Make advertisements specific and read the resumes you receive. If your advertisements are specific enough then you will not have too many resumes to deal with. You may choose to farm the more junior roles out to generalist agents if you do not have the bandwidth to deal with them but try to avoid putting key positions in your business out to general recruitment agencies. Utilize niche agencies where you feel comfortable and have worked with them before but be careful. Do not lose site of the importance of getting the best person for the role. Do not limit yourself to a local resource market; high achievers rarely stay in the one place for very long, they are in too high demand. Remember that the job market is now an active part of the global economy. Your aim should be to secure these people and do your best to hold on to them for as long as possible. For more junior roles by far the best approach is to hire graduates and train them in the role. This is cheap and you know what you are getting.

Essentially, what is required at this level of the model is that you have a hard look at the way your organization recruits staff. You may need to put a different process in place for key staff as opposed to non key staff.

4.5 Management Tools

The complexities of the functions that the continuity assurance team will be responsible for will mean that some management tools will be required to support the processes that the team implements, both internally within the team and across the wider organization.

These may include tools for:

- Risk management;
- Impact assessment;
- Problem management;
- Change management;

- Configuration management;
- Incident management;
- Knowledge base and case tools;
- Process analysis.

The Process management group should undertake to define the core operational processes of the team and determine what toolsets would be appropriate to support these processes. In extreme situations the development of a custom solution may be warranted. At this level of the model prospective tools should be reviewed and a strategic toolset selected. This toolset may need to be added to as more detailed requirements become evident.

Be aware that these tools are expensive. Change and configuration management alone could cost you several thousand U.S. dollars per seat.

4.6 Security and Safety

The principle security and safety activity at this level is to form the integrated security and safety team. This will not be a trivial activity. There will almost certainly be new recruitment that will need to take place in this area.

If possible and practical for your particular business, you should put in place a plan to discontinue the use of subcontracted security firms for physical security.

An information collation activity should be undertaken to ensure that information from the combining teams is brought together in a logical and ordered way.

An overall security and safety audit should be undertaken to determine the high-level gaps in the organization between physical security, IT security, and health and safety.

4.7 Level 1 Performance Indicators

It should be reiterated that one of the principal overall goals of this level, apart from the obvious creation of the organization, is the collation of information. It is important to have as much information about your organizations landscape as is possible. This will set you up for beginning the tasks at subsequent levels.

The level 1 performance indicators are set out in Table 4.2. Note that these are not performance indicators in the traditional or literal sense but more key tasks that need to be performed at this level of the model. How well you perform these tasks and how beneficial the outputs of the task are is what you will be rated on, in this way we refer to them as performance indicators:

Table 4.2
Level 1 Performance Indicators

Rationalization	
Item Number	**Action/Task**
R1-001	Create continuity assurance team and disband other teams with duplicate functions or merge them into the continuity assurance team structure.
R1-002	Sanitize and reorder overall organization structure to put functions in their logical place. Give particular attention to the existence, position, and function of the PMO within your organization.
R1-003	Ensure there are clearly defined communication pathways through the organization for key team interfaces.
R1-004	Implement a blended team for continuity assurance based on the hospital model, where appropriate (optional).
R1-005	Define and document team roles and responsibilities.
R1-006	Define and document internal management processes for the continuity assurance organization.
R1-007	Assess skills of team members and ensure that they align in terms of key skill requirements. Supplement with additional training where aptitude warrants the investment or move to replace.
R1-008	Assess prospective management tools such as problem management, incident management, configuration management, knowledge base, and case tools. Determine a strategic toolset to support the operational processes of the team. Acquire and implement these tools where appropriate.
R1-009	Gather general information pertaining to the continuity of the business and compile into a searchable repository.
R1-010	Commission projects were appropriate. These should pertain only to the implementation and establishment of the requirements under this level of the model (optional).
Management Functions	
Item Number	**Action/Task**
M-001	Develop communication road map for new initiatives and crisis management. This road map should address the communications required in transitioning the organization to a continuity centric model such as that laid out herein.
M-002	Define and implement a governance model suitable for business-as-usual and any projects that may eventuate.

Table 4.2 (continued)

Management Functions	
Item Number	**Action/Task**
M-003	Implement a system where problems may be identified, reported, and analyzed using root cause analysis. The system must be fully auditable. Management reports must be able to be produced in line with governance requirements.
M-004	Define and document high-level strategy including goals and timescales. This strategy may be further refined at subsequent levels. This strategy should cover a minimum of 3 years.
M-005	Set budgets based on the goals documented as part of the strategy.
M-006	Conduct an audit of overall security and safety to assess high-level gaps in the organization between physical security, IT security, and health and safety processes. Feed this information into the risk and vulnerability assessments that will take place at level 2.
M-007	Review high-level human resource processes with particular attention to recruitment. Form an action plan and initiate it.

Security and Safety	
Item Number	**Action/Task**
S1-001	Integrate physical security and IT security functions into the continuity assurance team.
S1-002	Put a plan in place to discontinue the use of subcontracted security firms, where possible. Initiate the implementation of this plan (optional).
S1-003	Review and assess all existing security procedures and security infrastructure.
S1-004	Review roles and responsibilities with regard to physical security personnel/guards.
S2-001	Integrate health and safety functions into the continuity assurance team.
S2-002	Review roles and responsibilities with regard to fire or safety wardens and building services personnel.
S2-003	Gather information on the following types of detail: Business locations and buildings (including schematics and environmental controls); Number of people at each location; Location of any people with disabilities; Configuration of escape exits.

5

Level 2: Risk Reduction

This level of the continuity assurance maturity model focuses on the identification and analysis of risks and the implementation of measures to reduce the probability of their eventuation. In addition, we will look at analysis of vulnerabilities within the organization, and likewise, the implementation of measures that remove or reduce these vulnerabilities.

Many of the requirements at this level of the maturity model will be around security and safety, as these are the defensive elements of the methodology and risk reduction is essentially a defensive exercise.

I should point out at this stage the difference between a risk and vulnerability. A vulnerability is a gap in your organization's infrastructure that may be exploited by a deliberate or accidental event. Vulnerabilities give rise to risks, being represented by the actual event that exploits the vulnerability. There are, however, many risks that are not based on vulnerabilities. The risk of flood for example is not based on any specific vulnerability within the organization. Because of your company's location you may be vulnerable to flooding but there is no specific vulnerability that could be identified or remedied, other than moving location, which we will talk about later. Improving a poor process or increasing security will not prevent the flood. You may implement measures to lessen the impact of the event but you cannot stop the event occurring by correcting a specific vulnerability. So, in effect, we can classify risks into two types. Those that are based on vulnerabilities that give rise to the risk itself, and those that are based on factors outside of your immediate control. Examples of such factors may be geographical or political environment, or weather. An example of a vulnerability could be a security hole in an IT system.

Given this symbiotic relationship between vulnerabilities and risks one should determine and address vulnerabilities before considering the more general risks to the organization.

5.1 Vulnerability Analysis

Key vulnerabilities will center on the following four areas:

1. Security;
2. Safety;
3. Process;
4. Architecture/design.

Weaknesses or gaps in security will normally constitute the most numerous and severe vulnerabilities. Inadequate safety measures and procedures will also give rise to specific vulnerabilities. An example may be inadequate organization in an evacuation situation leading to work being lost or security procedures not being followed. Criminals could exploit this vulnerability; indeed, it is not unknown for such evacuations to be deliberately initiated as a diversion for their exploits.

Gaps in process or ill-defined processes also give rise to vulnerabilities. We will discuss process in depth at level 4 of the model. Suffice to say that a gap in a key procurement process can bring downstream processes that are critical to the continuity of the business to a halt. I have seen business impacted severely just because someone forgot to place the order for printing paper or printer toner on time.

The final area of vulnerability is that which is inherent in the design of your organization itself, or in particular the infrastructure of your business. This could be a gap in communication borne out of a poorly designed organizational structure. It could also be integral flaws in the architecture of your IT systems, such as security holes or general application defects. On a network level you may have several ports open at your firewalls because you have an application that is designed to communicate over those ports; however, these open ports may give rise to a vulnerability in security terms. I once worked for a bank that spent a great deal of money on a piece of computer software that it was never able to implement because just such a vulnerability was discovered. An architectural vulnerability could also relate to physical architecture. The lack of an adequate fire escape would be both a security and architectural vulnerability.

You will find that many vulnerabilities can be related to security in some way. I recently worked in an office where there were only two doors in and out. A fingerprint access device controlled the main door. The other door was locked and the key had apparently been lost. The windows were barred and shatter-proof and we were on the second level. Unfortunately, the fingerprint device was on the main power grid and there was no backup supply. This fingerprint machine had to be used both to enter and exit the office. When the power went out, as it did one day, no one could exit the office until it came back on again. This is a good example of the balance that must be struck between security and sensibility. This is one reason why security and safety are grouped together under the CAM. Security must always be measured against safety, and what is sensible in business terms. The most secure premise would not allow any customers into the building, which clearly defeats the purpose of business continuity. The key word is balance.

At this level of the model you should be performing a detailed vulnerability analysis in terms of security, safety, business processes, and the underlying design of your organization and its infrastructure, both physical and virtual. These vulnerabilities must be identified, understood, and addressed before you can proceed with a general risk assessment. Remember that in resolving your vulnerabilities you may reduce the number of risks you have to mitigate.

Note that vulnerability analysis should be performed utilizing root cause analysis. It is the root cause that must be treated in order to truly remove the vulnerability.

5.2 Corrective and Preventative Action

You may ask how we can prevent a vulnerability from eventuating. We can achieve this by putting in place standards that ensure that any new implementation of system, process, or otherwise, is analyzed for vulnerabilities that could ultimately impact continuity at the time of initiation. The definition and implementation of such standards is a requirement at this level of the maturity model. These standards would also be utilized when assessing changes to existing systems, processes, and the like.

Corrective action should be conducted in a coordinated and controlled way. Where the change required is fundamental or the root cause complex you may need to take action over an extended period of time. If this is the case you may have a strategic solution and a tactical solution as previously discussed. The strategic solution will treat the root cause and may take some time to implement. The tactical solution will treat the surface problem and hopefully move the problem into a manageable state until the strategic solution becomes available.

Take care when making any change that you spend adequate time and effort in analyzing that change in terms of any adverse impact. The last thing you want to do is compound your problem or solve the problem and introduce a new one at the same time. We will talk in some detail about enterprise level change control processes in a later chapter.

You must be in a position where all identified vulnerabilities are resolved before moving on to your wider risk assessment. As I have mentioned, many risks exist because of the potential exploitation of a vulnerability. Address the vulnerabilities and you will have less risks to deal with.

5.3   Risk Identification

Risks may be grouped into the following categories:

1. Geographical/environmental;

2. Process failure;

3. Criminal;

4. Business operations.

The first represents natural events or disasters but also events based on the location of the business or business segment. Location is not just about your proximity to a fault line or flood plane; it is about how secure your premises can be made due to the landscape of the area. Your proximity to a particularly high-risk facility or lack of local access to emergency services could also be types of risk that are grouped within this category.

Process failure due to equipment breakdown, unavailability of resources, or accidental damage is a key category of risk. Note that this is slightly different than the vulnerabilities previously mentioned that are due to poor process design. Here there is no particular problem with the process itself, but there is always an inherent risk of failure due to an unforeseen event. There may be measures that can be taken to mitigate this risk such as implementing a level of resilience to high-risk systems. However, it should be noted that this is not seen as addressing a vulnerability, it is mitigating a risk. You may implement a failover system but you have not corrected the vulnerability. A vulnerability would be if there was an inherent problem with the design of a given system that meant that it often broke down. Altering the design so that this was no longer a problem would be to solve the vulnerability. That does not mean that the system will never break down again though.

Risks that can be categorized under the criminal area are, for example:

- Terrorist attack;

- Blackmail;

- Malicious damage;

- Theft;

- Trespassing with intent;

- Vandalism;

- Hacking.

A terrorist attack could be in the form of a direct attack, that is an event such as an explosion, fire, or biological agent. It could also be a hoax, threat, kidnapping, or hostage-type situation. You may hear the term CBRN used when referring to such incidents. This covers serious types of attack in the form of chemical agents, biological agents, radiological exposure, and nuclear attack.

Hacking can include the planting or propagation of viruses, denial-of-service attacks, or simply unauthorized access or unauthorized use of property or process. It is well to note that hackers are not always external to your organization. What I term "insider hacking" is actually much more common than the Hollywood version. This is, as the name implies, when someone within your organization accesses a system beyond the limit of his or her authority. Even if this person has what he or she feels is legitimate reason the person could be endangering the continuity of your business. Someone attempting to operate a system who has not been properly trained in the use of that system could make an error leading to an interruption of service. Of course the fact that they accessed the system in the first place means that there could be a vulnerability in the system. Additionally, in this sort of case, and in line with our principle of root cause analysis, we need to look at why the person accessed the system. In many cases you will find that this is a shortcoming in some management process. For example, a small IT call-center may have 20 operators and one supervisor. It may be the case that when a user requires a password reset the supervisor has to access the system where this task is performed, to authorize the reset. Because of the volume of password resets the supervisor may distribute his or her login details to the operators so that the process works quickly and smoothly for the users. This is very pragmatic of the supervisor but points to a clear problem with the password-reset process. On the surface this may look good for business but could prove disastrous should something go wrong.

The mitigation of all of these crime-related risks could be looked at in terms of security, safety, or process.

Moving back to our categories of risk we finish with what I have termed business operations. These are often not true risks in continuity terms and should only be considered so in the most extreme circumstances. Here we are talking about trading risks such as fraud, revenue leakage, inflation, currency risks, and interest-rate fluctuations. We will not mention these again as they are not the subject of this work but it is likely that some of these, fraud for example, may be mitigated through the implementation of this methodology.

5.4 Risk Mitigation

From the categories of risk previously discussed we will look at ways to mitigate risks in the area of:

- Location;
- Environment;
- Fault/failure [of process or supporting system(s)];
- Security;
- Safety.

It should be noted that the selection of business location(s) should be subject to assessment against the standards already mentioned, whose definition is a requirement at this level of the maturity model. For existing locations, any vulnerabilities relating to the location should be assessed and addressed. Where there is a particular risk associated with the location you should put in measures to attempt to mitigate or lessen this risk.

Where business objectives and practicalities allow, the permanent move to an alternate location should always be an option. A "greenfield" site will generally be easier to secure than a city location. This is due to the proximity to other buildings or businesses sharing your building. The ideal situation is the ability to create a physical perimeter around your site. This prevents anyone gaining easy access to your building, as the first level of security is located away from the building itself. With such a location you are able to create clearly visible entry points that can be well monitored and controlled. Entry and exit of vehicles is generally more controllable than individuals, though it can be time consuming. Again, balance is the key. If we are rigorous to the point of impracticality for the operation of the business then the rules are likely to be bent and security weakened to a point below that reached by a more pragmatic approach to begin with.

A measure of geographical diversity can reduce risks associated with location. This basically follows the principle of not having all your eggs in one basket. If your organization is geographically diverse enough you may be able to withstand the loss of one or more sites. This clearly lessens overall location-related risk to your business.

It is sensible to minimize the traffic of individuals and vehicles into and out of your premises by providing as many facilities within the complex as possible. These may include eating and sporting facilities as well as a range of other services. Many organizations now have coffee shop franchises and travel agents located within their buildings. Some even have doctors and child care facilities.

Visitor management is a key risk area and there are various methods of managing this. As an example, visitors should always be given ID tags and should have verifiable prebooked appointments. Visitors should be seen in meeting rooms and should not have access to the general working areas of the office. Premises should be set up to ensure that there are adequate meeting facilities to cope with requirements such as this.

All packages should be delivered via a central mail sorting area. This mailroom should always be located away from the primary business premises. All mail should be scanned and/or preopened.

Offices should be designed with security and safety in mind. For example, offices should not be placed in the same location as your data center, if possible. Smoke sensors and the like are standard and are generally covered by law but companies should consider monitoring of other types of threat. It is possible to obtain sensors for toxic chemicals that work in a similar way to smoke detectors. It is not unheard of for companies to make extreme alterations to their building designs to take account of security and safety. To this day new buildings in Zurich, Switzerland, are built with bomb shelters incorporated into the design. This is actually a regulatory building requirement in Switzerland.

If you have the option, do not locate in an area that is known for environmental problems. Move away from areas that suffer from earthquakes, flash floods, bush fires, and the like. Where you cannot move ensure that you have adequate measures in place to deal with such events. Simple measures like having a warehouse full of sandbags stored for use in a flooding emergency can be invaluable. In some cases you may need to look at provisioning your own emergency services. Do not rely on the fact that the state services will be available to you in a timely manner if your area is suffering a general disaster. Airports do not rely on local fire brigades to deal with their emergencies, they provision their own to ensure they always have some level of immediate response capability. If you are in an area with a high fire risk make sure you have additional extinguishers, incorporate fireproof walls in your building design, and build with fire-resistant materials if possible. Make sure you do not make any modifications to your building that negate any design measures that have already been thought of and implemented. It is not disaster related, but how many times have you been in an office that was freezing down one end and hot at the other. This is almost always because someone has erected a wall or partition where they should not have. The air conditioning systems would have been designed based on the airflow in an open-plan space. If you erect a wall down the middle of the space you will get problems such as this. Think about what you are doing. Assess the impact on the comfort, security, and safety of your personnel. Every change should go through a formal change management process, the key purpose of which is impact assessment.

Ensure that processes are well defined, well understood, and well communicated. Ensure that people using supporting systems are well trained in their use and that there are processes in place for use when something goes wrong. Identify and address vulnerabilities in these processes and mitigate any residual risks. At this level of the methodology we are looking at the processes that exist and refining these to lessen risk. There may be process gaps identified for which new procedures will need to be defined. We are not yet at the stage of implementing any specific physical capabilities in terms of resilience or recovery. At this level we are assessing existing processes and fine tuning. This fine tuning will form a tactical response to the problems we identify. Subsequent levels of the model will provide the more detailed and complete information we will require to plan and implement strategic solutions.

5.5   Security and Safety

In level 1 of the CAM we integrated physical security, IT security, and health and safety. At this level we will leverage this integrated structure to improve general security and safety across the organization. This level requires the assessment of all security and safety procedures and systems, and the definition of both tactical and, where possible, strategic solutions, in parallel, to any problems or inadequacies identified. Implementation of these solutions should also occur at this level. We have seen that the majority of risks can be reduced by the implementation of specific security and safety measures. As this module is about risk reduction, security and safety measures constitute the bulk of the work to be undertaken before moving to level 3 of the model.

Below is a list of miscellaneous points that may be useful in improving overall security and safety. Most are reasonably obvious, some a little less so.

- All physical access points should be constructed using an exclusion zone. An exclusion zone is formed between two barriers. Only one individual or vehicle is allowed in the zone at any one time. This allows the individual or vehicle to be assessed and prevents "tailing," where one person follows another without having to be assessed separately.

Where vehicles are concerned the second barrier should be constructed in a way such that it is sufficient to stop a vehicle attempting to ram it.

- Where appropriate, vehicles should be checked underneath using mirrors, in the engine bay and a cursory look inside for any obviously suspicious packages. Engines should be stopped while this check is being undertaken.

- Vehicles should be identifiable via issued ID tags. Taxis that frequent the site should also be registered, both driver and vehicle. The same applies to delivery vans. If possible have your own cars or minibuses and do not use taxis at all.

- Ensure that system backup tapes are stored away from the primary site and make sure that they are transported securely. Remember that these tapes contain critical and sensitive business data.

- Security passes should be used in conjunction with some personal identifier such as PIN number, fingerprint, or the like. This is to prevent passes being used if stolen. Your building should be divided into a number of secure areas with individuals given specific access only to the areas that they need to have access to.

- Do not allow any cameras in your premises other than those authorized for business use. Yes, this may include those in mobile phones.

- Ensure that security and emergency staff have secure communication such as VHF or UHF radios. In a general disaster mobile telephone networks can become overloaded and rendered unreliable. Pagers are often also a good alternative as they operate on a separate telecommunications network to mobile telephones. In extreme locations, access to a satellite telephone is also useful.

- Consider security in transport for key personnel. Clearly this depends a little on your location and any specific risks associated with your business. Consider restricting the number of personnel that can travel on any one flight. This is to minimize impact to the organization should an event occur that affects the flight.

- Where possible, implement an inoculation plan for your employees, particularly if they have to travel abroad for business reasons. Something as simple as a flu vaccination can help to assure continuity. Family medical plans can also have a consequential effect in reducing continuity risks.

- Ensure that security and safety inductions are carried out immediately for all new personnel, including temporary staff. This also applies to existing staff moving to a new building. Ensure someone is responsible for the safety of visitors, trades people, and the like. Such people should not be left unattended.

- Do not allow any food and drink in technically intensive areas (e.g., data centers).

- Make sure that evacuation meeting points are well thought out and sensible. In recent terror attacks on residential compounds in Saudi Arabia terrorists ran through the compounds after the bombs had exploded shooting at anyone they could find. In one compound I stayed at the meeting point was by the pool, the most open space in the compound. This was obviously changed as it was deemed safer to stay inside the buildings until told that it was safe.

- Ensure that all processes and procedures are regularly tested, where possible by an independent external agency. Conduct active probing of security where appropriate, including penetration testing and other intrusive testing. Make sure this is well planned and conducted at a low-risk time. We do not want to impact business continuity with a failed test of continuity assurance capabilities.

When implementing security measures please be aware of people's rights in terms of civil liberties. This is particularly relevant, but not limited to, government agencies, airport authorities, and anyone implementing security measures that are intended to be applied to the general public. Be careful that you do not implement something with the best of intentions to solve a specific problem and then end up using that solution or technology for other means. For example, do not put in cameras for security and use them to see how long people spend at their desks. Recent events call to mind American plans to fingerprint foreign visitors and plans to include fingerprints on passports. One has to wonder where this will all end. Again, balance is the key to usable, effective security. You may put people to a lot of pain and inconvenience and find that you have not improved their security or safety situation at all. Remember that agitation can create a situation that would have otherwise not occurred. At all times ensure that the subject of your security procedure is treated fairly and politely.

I feel compelled to mention an incident that I witnessed, again in Los Angeles. Security personnel were searching all hand baggage upon boarding the flight that I was traveling on to Auckland, New Zealand. They were also searching people and checking shoes, for everyone, including children. They had a steel bench set up just between the boarding counter and the door to the access bridge. A family with a baby and a number of bags was struggling with this process and had placed the child on the end of the table while they attended to the bags. This was a fairly long table with two or three security personnel checking bags in parallel. People were getting backed up as this family struggled and they had left a baby bag at the other end of the bench by mistake. One of the security staff, annoyed at this delay ripped through the bag and then slid it hard across the bench into the baby sitting at the other end knocking the baby clear off the bench. No apology was offered and the argument that ensued almost resulted in the father of the child being arrested. The only statement made by the security officer, "Don't be annoyed with us, we are doing this so that you are safe!" Well, I did not feel safe around these people and I do not think the baby felt too safe either. The message, be careful what you do and how you implement it. A good idea on paper can turn into a disaster when implemented for real. If you do not want to look stupid, treat security as a public relations challenge. Do not put people without public relations or management skills in this type of role. Make sure there is a relatively senior person on hand to deal with any problems that may eventuate.

5.6 Level 2 Performance Indicators

Table 5.1 sets out the performance indicators at this level of the continuity assurance model.

Table 5.1
Level 2 Performance Indicators

Risk Reduction	
Item Number	**Action/Task**
R2-001	Conduct a detailed vulnerability analysis of business processes and systems. Ensure that the root cause of any vulnerability is identified.
R2-002	Define a standard and process for the assessment of vulnerabilities when implementing any new system, process, or architecture. Equally, apply this standard to assessment of changes to existing systems, process, or architecture. Implement this standard and process.
R2-003	Resolve or control all identified vulnerabilities by implementing tactical or strategic solutions. Where tactical measures are taken a strategic solution must be planned or in work.
R2-004	Following the resolution or control of vulnerabilities conduct a general risk assessment across all areas of your organization. You may preclude operational risks such as fraud, revenue leakage, and the like except where you consider that such problems could impact general continuity.
R2-005	Assess risks in terms of location, environment, crime, and process, and plan any mitigating action. Implement these measures.
R2-006	Compile a list of residual risks that will need to be considered at later stages.

Table 5.1 (continued)

Management Functions	
Item Number	**Action/Task**
M-008	Ensure process changes are well communicated to all parties involved in the process. Coordinate additional training where required.
M-009	Ensure that the identification of risks and vulnerabilities, and the resolutions actioned, are documented in an auditable fashion and added to the department's information repositories.
M-010	Foster a culture where risks are thought about day-to-day and implement a mechanism whereby any members of staff can communicate any risks they feel they have identified.
M-011	Scope, commission, and govern projects where required that are borne out of the fulfillment of objectives.

Security and Safety	
Item Number	**Action/Task**
S1-005	Conduct a detailed vulnerability analysis of security processes and systems. Ensure that the root cause of any vulnerability is identified. Implement resolutions where possible.
S1-006	Review and refine security standards and the processes that facilitate the implementation of these standards.
S1-007	Assess risks in terms of security and plan any mitigating action. Implement these measures. Add residual risks to register.
S2-004	Conduct a detailed vulnerability analysis of safety processes and systems. Ensure that the root cause of any vulnerability is identified. Implement resolutions where possible.
S2-005	Review and refine safety standards and the processes that facilitate the implementation of these standards.
S2-006	Assess risks in terms of safety and plan any mitigating action. Implement these measures. Add residual risks to register.

6

Level 3: Rating

At this level of the continuity assurance maturity model we are concerned with the rating of components of your organization's infrastructure in terms of their criticality to the continuity of the business. We will determine this by assessing the impact to the business by the unavailability of each of these components.

Clearly, it is not possible to assess every component of the organization so we constrain our assessment to components that are utilized in critical business processes. The assumption is that if a component is not part of a critical process then it cannot be critical to the continuity of the business. A critical process is defined as a process that if stopped will impact continuity in a way which is unacceptable. What is acceptable and what is not is something that your organization will have to decide for itself.

The exercise of assessing business processes and the components of the organization that support those processes for business criticality is called business impact assessment (BIA).

A component could be a system, functional role (individual), piece of equipment, place, interface, or connection.

Once we have determined the impact of loss we refer back to our residual risks to ascertain the probability of the loss. These two factors of impact and probability will be jointly utilized to identify components that need to be given some attention at subsequent levels of this model.

6.1 Business Impact Assessment

This activity is undertaken in much the same way is it is done under traditional business continuity methods, except for the fact that our definition of a component is a little broader than most.

We begin by looking at the business in terms of the products and services that the business offers. We will collectively call these the products, even though some may indeed be services. Once we have identified the products we look at identifying the key business processes that enable those products to be provided. At this stage it is useful to have some idea about the particular business you are assessing. Most businesses have a standard process model that can be used as a starting point. I have already used as an example the standard process model for a telecommunications company. Banking, insurance, manufacturing, and so on all have similar standard operating models. These define the high-level processes at work within the organization and may include such things as sales, procurement, and customer service.

If we start with the products of the business we may look at these in terms of criticality. What would be the impact if we could not supply product X? You may find that you can eliminate a whole product line from the beginning because this product does not have a sufficient customer base or does not earn sufficient revenue for it to be critical to the business. This does sometimes occur where a company is prototyping a service or is dabbling in something outside of their core business.

It should be noted that there are many factors that need to be considered in terms of impact to the business. These are the continuity indicators that we have discussed previously and are generally accepted to consist of the following factors:

- Financial;
- Reputation;
- Customer service;
- National security;
- Health and safety;
- Regulatory.

Hence, we look at product X and decide the impact of not being able to provide this product in terms of each of these six indicators. For example, a product that you are merely prototyping may have near-zero impact in terms of finance or customer service but the loss of this product may be unacceptably detrimental to your reputation, perhaps because of marketing activities that have already been initiated.

Once we have assessed all products we can move on to the high-level processes that we have obtained from either our business model or our analysis efforts within the organization. We may preclude some of these high-level processes without having to go to any further level of detail. If the organization keeps large stores of stock and supplies on hand then the procurement processes, for example, may not be critical to its short-term continuity.

For the products and enabling processes that are deemed critical at this level, we need to perform more detailed analysis in order to reduce these to their constituent parts and determine which are the critical components that enable the parent process, and thus the product.

We must reduce each process to its subprocesses and assess the criticality of each subprocess in turn. At each level, for those that are still deemed critical, we must take the analysis down to the next level, and then the next, and so on.

For the remaining low-level critical processes we then look at each of the components that enable that process to function. These components will consist of:

- Equipment/machinery;
- IT systems;
- People;
- Place/environment;
- Information/data;
- Communication network(s).

We should then have a complete criticality map in the following layers:

1. Products;
2. High-level business processes;
3. Low-level business processes;
4. *n*-level business processes;
5. Components;
6. *n*-level subcomponents.

It should be possible to see from this criticality map which components are critical to multiple processes, in a similar way to that illustrated in Figure 6.1.

It should be noted that the criticality analysis at each process level, and of each component at this stage, is performed in a qualitative manner. We will discuss further quantitative analysis and classification later in this chapter and the next.

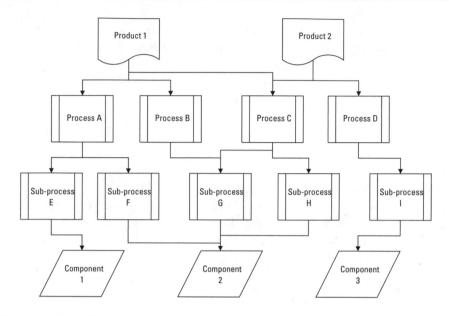

Figure 6.1 Criticality mapping.

It should be determined at this stage what the business tolerances are around the failure of any of these processes based on the customer's tolerance to the unavailability of the product or the impact to the business in terms of any of the continuity indicators. These tolerances should be expressed in terms of time and loss of information/data. It should further be documented if there are any resilience or recovery capabilities currently in place for the components, or if the processes have available any proven manual workaround. That is, a workaround that does not require the use of some or all of its dependent components. This information will later be utilized to determine recovery time and recovery point objectives.

I think it is useful to look at an example of a typical process and its components. Let us look at something simple like a pizza shop taking a delivery order. The total process map may look something like Figure 6.2.

But what is critical and what is not? The following points provide some insight:

- The telephone and telephone network are critical, an order cannot be received without them.
- The ordering system can be bypassed and orders taken manually. It is therefore noncritical.
- The sales assistant is not a specialist role so we do not consider it critical.

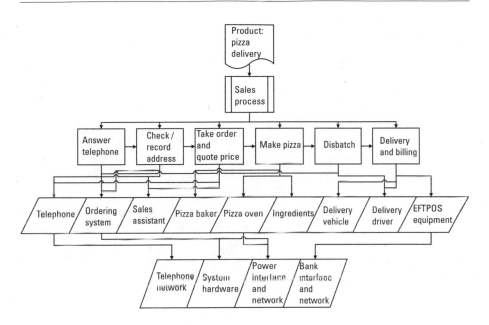

Figure 6.2 Example of a criticality map.

- The pizza baker is a specialist role and is considered critical.

- The pizza oven is critical, the product cannot be produced without it.

- Some ingredients will be critical, such as the ones that are used in the majority of orders. There may be some customer tolerance to some ingredients not being available. For example, pepperoni and olives may be critical but anchovies may not be.

- The delivery vehicle I think is critical, but could be replaced by the use of a taxi. It would not be cost effective but the reputation of the business may remain in tact.

- The delivery driver is not a specialist role and is therefore not critical, the role can be easily replaced or another member of staff can perform the role.

- The EFTPOS or point-of-sale equipment that processes your card payments will impact customer service as, if it was unavailable, only cash payments could be accepted. It would need to be determined by management if this was an acceptable impact or not. I would say yes, but for a limited time only. The component is critical but the recovery-time objective is likely to be short.

Note that components can also have subcomponents that need to be included as part of this analysis exercise. It is possible that only some subcomponents of a system are considered critical. For example, you may have a financial support system which consists of several financial modules. Each module may be utilized to facilitate entirely separate business processes. Some of these processes may be critical and some may not. A payroll process may use a payroll module of the system and hence that module would be critical, within a given tolerance. Depending on the architecture of the system it does not necessarily follow that the entire system has to be looked at in terms of some robustness, resilience, or recovery capability given that only part of the system is critical. The part of the system that is critical will still need to be provided for though.

As in the example above, when we are considering IT systems we always look at the software layer first, and then define any hardware that facilitates the use of the software subsequently as a subcomponent of that system. Where we have an *n*-tier application architecture we look at each software layer in turn before moving onto the hardware components. Similarly IT networks form another layer underneath general hardware, that is, servers and server components.

At this point we have not defined what "critical" is in any specific terms. The aim of the BIA is to document a range of impacts across all business processes. Once we have this range, in largely qualitative terms, we can draw a line under what is acceptable impact and what is not. Where we have said this is critical and that is not in the example above, we have made a level of judgment on the perceived impact of a process being stopped. It is a simple example and you will need to go into some detail when you do this for real but I believe it demonstrates the general approach. As previously mentioned, if you want to introduce a quantitative perspective you should look at impact in terms of recovery tolerance. You should also look at consequential effects. If you have a low-level process that is used within a large number of higher level processes then the impact of losing this process is quite widespread. You must look at the effect of this on the other related processes. Be careful not to look just at downstream impact. The failure of a downstream process equally may have an impact on upstream processes.

Once we have determined a range of impacts for our organization we are in a position to apply a rating model for criticality.

The BIA should be compiled as a document and should include a criticality map like the one above. This report is a requirement at this level of the continuity assurance maturity model.

6.2 Criticality Rating

There are varying degrees of criticality. Saying that a component is critical to a process does not fully define the nature of the criticality.

Note that the BIA is conducted in a top-down fashion. The idea is that by assessing top-level processes you will eventually drill down to the individual components and subcomponents that support the underlying processes. This is true for all types of components other than IT systems. We have talked at some length about the disconnect between the business and IT within many organizations. The larger and more complex the organization the worse this disconnect often is. The result is that you may not always obtain clear and complete information about IT systems in conducting your BIA. Some process owners do not really have a complete understanding of the IT infrastructure that enables their process. Furthermore it is useful to have a complete picture of the IT infrastructure of your organization as a whole and is thus useful to rate each and every system in terms of criticality. So as part of our criticality rating process we look at each IT system bottom-up and assess the criticality of each (see Figure 6.3).

This process of systems criticality assessment (SCA) should be performed in parallel to the BIA and information should be shared between the teams performing each task. In this way you will be sure to build up a complete and accurate picture of your organization's critical components. Systems criticality assessment is begun from within IT and works up to the business layers; and the BIA begins in the business layers and works down towards the IT areas.

We can classify criticality of processes and components in the following way; see Table 6.1:

You will need to define what you consider to be high, medium, and low in terms of the impact information and range determined from the BIA.

High may mean that a level-one process is stopped completely. Medium may be reflected by the stoppage of a second-level process, which impacts the

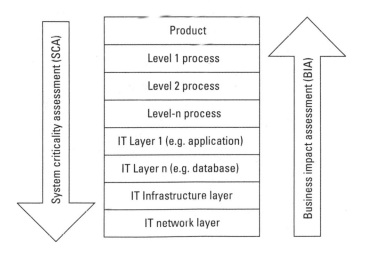

Figure 6.3 Criticality assessment method.

Table 6.1
Criticality Classification

Criticality Classification	Description
Primary	High impact to continuity indicators.
Secondary	Medium impact to continuity indicators.
Ancillary	Low impact to continuity indicators.
Noncritical	No impact to continuity indicators.

completeness of a level-one process. Low may mean that a low-level process has stopped and there is a minor impact or delay to the functioning of a higher level process. Non-critical processes or systems are those that have no unacceptable impact to any part of the business in terms of the continuity indicators. Primary, secondary, and ancillary processes and components all represent a level of unacceptable impact in terms of the continuity indicators.

When looking at IT systems specifically it may be useful to chart criticality against any known recovery capability. Note that multiple systems are plotted on a common chart (see Figure 6.4).

In general terms the recovery time for any of the systems charted should fall within the shaded square at their particular criticality level. If a system over performs at its level then this is fine, but it may indicate an "overspend" on

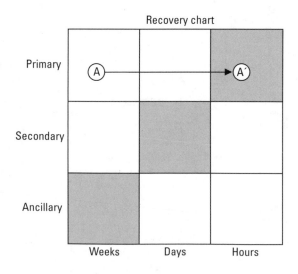

Figure 6.4 Recovery mapping.

cap rrent observed cap ty and "A[1]" the desired
cap understanding of mer tolerance or busi-
nes wo quite clearly re ts the capability gap.

gr the chart represen ero at the low end and
th you may wish to order
le least critical within the

i

y known resilience built

nd "B[1]" the desired capa-
ience capability warranted

lity for IT systems.
ine if that system meets a set
eria may be derived from any
ermined from your particular
systems that fall within a par-
ion. Systems may be excluded

t change which should be com

not part of the core operational

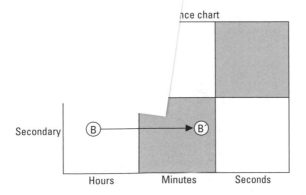

Figure 6.5 Resilience mapping.

To quantify these, an assessment form may begin something like that in Table 6.2.

Each criterion should be weighted in terms of importance. The weighting puts more importance on some answers than others. If a system is not in production then it is very unlikely to be critical; however, a system may be scheduled for retirement but this may take longer to happen than may be planned. While the system is in production it can still impact the continuity of your business. The form should imply a yes or no answer depending on a score threshold that you will need to set. Note that all questions here are binary, yes or no.

The following form requires that you first understand and set out the following:

- List your high-level business processes (these are the level-one processes previously discussed);
- List your products or service areas;
- List the key customer segments that your organization serves.

For each of these you will need to understand their relative importance in terms of your business and the continuity indicators. You will need to weight each in terms of most-to-least important. The most important should be weighted with the highest number. You may have some products that are more important to your business than others and some types of customer that you consider more important than others. Table 6.3 sets out a framework that can be used to determine a provisional criticality rating.

Please note that the weightings shown here are for example purposes only. Again, you will need to decide what importance you put on each of the continuity indicators in order to determine these weightings for yourself.

Clearly, in terms of our pizzeria example, national security is not a factor. Regulatory reporting, however, could encompass recordkeeping for tax purposes as well as employee-related regulatory requirements.

Table 6.2
Preassessment Criteria

Preassessment Criteria	Guide	Weight	Answer	Rating
Is the system currently in production?	0 or 1	3		
Is the system stable in production?	0 or 1	2		
Is the system scheduled for retirement?	0 or 1	1		
Is the system eligible for assessment?				Yes/No

Table 6.3

Provisional Criticality Assessment

Criticality Assessment	Guide	Weight	Answer	Rating
What key business process does the system support?				
Sales	0 or 1	3		
Billing	0 or 1	2		
Logistics	0 or 1	1		
What product area does the system support?				
Restaurant	0 or 1	3		
Delivery	0 or 1	1		
Drive-through	0 or 1	2		
Who are the customers of the products that this system supports?				
Families	0 or 1	3		
School children	0 or 1	1		
Single adults	0 or 1	2		
Rate the impact of loss of this system on the product area(s)?	1 to 5	5		
How many customers per month does the system serve?	1 to 5	4		
What is the rate of increase of customers for the product(s)?	1 to 5	3		
Rate the projected future growth of the product area(s)?	1 to 5	4		
How many users does this system serve?	1 to 5	3		
What is the monthly revenue generated by the product(s)?	1 to 5	8		
Would the impact on the product(s) result in any impact to national security?	0 or 1	50		
Rate the impact of unavailability on reputation.	1 to 5	7		
Does the system support any regulatory reporting requirements?	0 or 1	30		
Would impact result in any safety issue?	0 or 1	40		
Rate the impact of unavailability on security.	0 to 5	10		
Rate the impact of unavailability on the customer.	1 to 5	9		
Criticality Score:				Max: 403
Provisional Criticality Rating (1):				Primary?

This form may give a total criticality score out of a maximum of 403. You can now divide this score by arbitrary cut-off figures or ranges for each level of criticality. For example, noncritical may be below 100, ancillary from 100 to 199, secondary from 200 to 299, and primary 300 and above. Again, these figures are for example purposes only. This will give you what I call a provisional criticality rating.

There are two more factors that we must consider before we decide on the final criticality rating of the system. These are:

- Recovery time objective (RTO);
- Recovery point objective (RPO).

The RTO should be evident from the information contained within the BIA. This is the figure represented by the customer's or business's tolerance to the stoppage of a key process. If the system supports this process directly or indirectly the system will inherit its recovery time to some degree, depending on the exact nature of the relationship.

The RPO describes the customer's or business's tolerance to data loss. The recovery point is the point at which data is recovered to. A system may be unavailable for 1 hour and when recovered may only have the data current to the last system backup, which may have been 12 hours ago. Hence the recovery time is 1 hour and the recovery point is 12 hours.

You will need to determine the RTO and RPO for each system under assessment. The following forms represented by Tables 6.4 and 6.5 may be useful in quantifying these. The RTO and RPO may have an influence on the

Table 6.4
Recovery Time Objective Assessment

Recovery Time Objective (RTO)	Guide	Weight	Answer	Rating
Rate the impact of an outage of the following durations.				
	1 hour	0 to 5	30	
	4 hours	0 to 5	25	
	8 hours	0 to 5	20	
	12 hours	0 to 5	15	
	24 hours	0 to 5	10	
	> 48 hours	0 to 5	5	
RTO Score:				Max: 525
RTO:				4 hours?
Revised Criticality Rating (2):				Primary?

Table 6.5
Recovery Point Objective Assessment

Recovery Point Objective (RPO)	Guide	Weight	Answer	Rating
Rate the impact of *not* achieving the following in event of recovery.				
Nil loss, even throughout outage.	0 to 5	35		
Recoverable to point of failure (PoF).	0 to 5	30		
Recoverable to PoF with inconsistencies.	0 to 5	25		
Recoverable to 8 hours prior to failure.	0 to 5	20		
Recoverable to 12 hours prior to failure.	0 to 5	15		
Recoverable to 24 hours prior to failure.	0 to 5	10		
Recoverable to last available backup (>48 hours)	0 to 5	5		
RPO Score:				Max: 700
RPO:				<8 hours?
Revised Criticality Rating (3):				Primary?

criticality rating. Your provisional criticality rating may rate the system as secondary but if we show that there is no tolerance for data loss then the system will probably be primary. Remember that our criticality rating will drive the capabilities we put in place to assure continuity. A system that has zero data loss as a requirement requires the highest level of capability in place to achieve this.

You can assess the RTO by looking at the outage time at which the situation becomes unacceptable. As a guide, the first time period to hit a rating of 4 or above should give an indication of the RTO.

You may define rules such as any system with an RTO of less than 1 hour will be a primary system. Less than 8 hours will be secondary. Less than 24 hours will be ancillary, and more than 48 hours will be noncritical. Using these rules we can utilize this RTO assessment to upgrade systems where required to a higher level of criticality. This method should not be used to reduce criticality level. Similarly RPO can be assessed as follows:

This information can be utilized to further revise your system criticality rating and give you an idea of what your RPO is. As a rule, any system that has a requirement on it of nil data loss, even throughout the outage, should be categorized as a primary system.

There are other questions that you can include in these forms to further refine your rating process. These may be, for example:

1. Is the data on the system replicated in any other system?
2. How often is the data on the system backed up?
3. Can the data be easily reconstructed from other data in other systems?

6.4 Risk Rating

Now that we have a good idea of criticality we need to weigh this against the work we have previously done on risk assessment.

We should have a list of residual risks as output from the last level of the continuity assurance maturity model. We now need to see where the impact of an event has an effect on the critical components of our business.

We should have enough information to compile a relationship diagram with risks on one side and critical components on the other. Consider the example of an airline company in Figure 6.6.

Once we understand these relationships we can further rate the impact of the risk eventuating. Remember that we have already implemented some measures to reduce or eliminate the risk, the risks that are left we call the residual risks. It is possible that some of these residual risks do not affect any critical component, hence that risk can be dismissed at this level and no further action is required at this time concerning that particular risk.

We can rate risks in terms of two factors. These factors are:

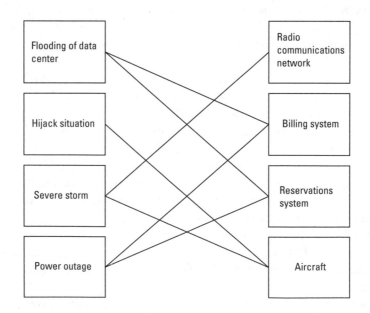

Figure 6.6 Relational mapping.

1. Probability;

2. Criticality of components that are impacted.

We can utilize the map in Figure 6.7 to illustrate this. We chart probability of the risk eventuating against the severity of the impact, in terms of which components are impacted.

In this way we draw attention to what we will call the critical risks, which are basically any risks that fall within the shaded area of the graph. These risks need to be further mitigated through the increase in defensive or recovery capabilities of the critical systems that they could impact. This is not to say that other risks are noncritical, they are just less critical, but may still require a level of attention. The purpose of this exercise is one of ranking and priority, and of increasing awareness of the critical risks.

6.5 Security and Safety

Security and safety are considered in terms of process and components just as any other areas of the business and are included in the assessments above. You will note from the criticality assessment form example, that security and safety are considered within the assessment of each and every component.

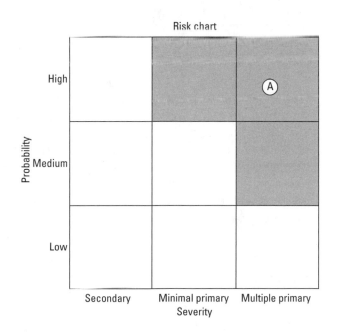

Figure 6.7 Risk mapping.

There are no additional security or safety requirements at this level of the continuity assurance maturity model.

6.6 Level 3 Performance Indicators

Table 6.6 sets out the performance indicators at this level of the CAM.

Table 6.6
Level 3 Performance Indicators

Rating	
Item Number	**Action/Task**
R3-001	Conduct and document a full BIA across the whole organization.
R3-002	Conduct and document a systems criticality rating (SCR) for all IT systems using both qualitative and quantitative methods.
R3-003	Produce a recovery capability chart and a resilience capability chart for IT systems.
R3-004	Compile a relationship diagram between residual risks and critical components.
R3-005	Produce a risk chart based on risk probability and severity of impact. Identify critical risks.
Management Functions	
Item Number	**Action/Task**
M-012	Socialize within the wider organization all results of activities within this level as and when they become available. Ensure feedback is relayed to the functional teams for consideration and possible revision/incorporation.
M-013	Ensure quality of process by conducting a quality review at some point within this level. Ensure that methods used adhere to the methodology and are consistent across activities. Produce a report on your findings and implement corrective action where required.
M-014	Ensure that documentation is compiled in an auditable fashion and added to the department's information repositories.
M-015	Continue governance of projects commissioned at earlier levels.
Security and Safety	
Item Number	**Action/Task**
S1-008	Assess critical risks and determine which are security related.
S2-006	Assess critical risks and determine which are safety related.

7

Level 4: Rigor

At this level of the continuity assurance maturity model we are concerned with rigor of process. There are certain processes that will need to be created or strengthened in order to prepare the organization for subsequent capability implementation. Resilience and recovery capabilities need a firm process base to be built on. You may implement a technical solution but you only truly have the capability to utilize it when you have the processes in place to manage and execute the solution.

It is important to foresee the processes required and implement these processes early. As we move on to define and implement technical capabilities we may need to further refine these processes and indeed refine them constantly over time. Processes can be tweaked at a later date to reflect the exact nature of the capabilities implemented. Conversely, defining the processes at this level may highlight issues or considerations that need to be assessed when designing technical capabilities and may have an influence on the overall solution architecture.

Note that this level of process refinement is different to that undertaken at previous levels of the model. We have talked about the rationalization of existing operational processes within the organization and reduction of risk, in some cases through process refinement. At this level we are looking specifically at processes that require a new level of sophistication, or indeed creation, to prepare the landscape of the organization for the introduction of new continuity assurance capabilities that are to come in subsequent levels of the model.

Here we will discuss a number of process areas that are essential for most continuity assurance capabilities to be usable and successful in application. These are:

1. Change and configuration management;
2. Contingency planning;
3. Crisis management;
4. Design assurance;
5. Backup and tape management.

We have said before that this methodology is designed to be followed in sequence. Many will argue that they need some immediate technical capabilities and cannot afford the time taken to follow this methodology. Remember that this methodology does provide for the implementation of tactical solutions at any time on any level within the model. However, you cannot be accredited with achievement at any given level until you have met all the requirements of the preceding levels and revised your outputs based on any new information coming out of such a backtrack through the levels. The reason for this can be best illustrated by an example.

The purpose of the graded approach can be seen when we consider a process such as change management. Very simply, if we have an IT production environment that is managed through a poor change control process then we never really have an exact picture of what the environment looks like. If we are also running a recovery environment for use in event of disaster then this environment needs to mirror the critical components of the production environment exactly, at least in terms of the application layer. If a change is made to the application in production and that change is not immediately replicated in the recovery environment then we have a problem with the synchronicity between our environments. If we have a disaster situation at that point and need to move operations to the recovery environment then we may find that elements of this environment do not work as they should. Say that we have made a change to a database table that replaces one field and adds another. If this field change is not replicated in the recovery environment then we will be trying to recovery data from production to tables in the recovery system that simply do not match. This will create all sorts of data inconsistency problems and may ultimately lead to a totally ineffective recovery capability.

The message is that without taking a sequential approach to continuity assurance you may find that you spend a lot of money implementing a capability, say a recovery capability, that may be workable at the time you commission it, but is relatively useless 5 minutes after. This sort of event happens when an organization is not adequately prepared for the capability that they are implementing. Continuity assurance capabilities will often alter the fundamental landscape of your organization. These capabilities must be implemented in a structured and coordinated fashion. There may be a large number of changes that need to occur in all parts of the organization for this capability to be truly achieved. These changes are all predictable with the right level of analysis effort and should be addressed prior to, or at the same time as, the implementation of the capability. In the case of process changes, these need to be made well ahead of time to ensure that the process is adequately "bedded-in," that is, tested, communicated, and so forth, before the capability is introduced.

7.1 Change and Configuration Management

The importance of good change and configuration management cannot be understated. Unfortunately change and configuration management is not well understood and not well done in most organizations. As a result, capabilities such as disaster recovery, from a technical recovery perspective, are not well implemented in most organizations.

You will note that I have grouped change and configuration together. This is only for the purpose of understanding because most organizations consider the two areas separate. This is in fact a gross misunderstanding as one is actually a subset of the other. Change management is actually a function within configuration management.

Everyone has a different opinion of what configuration management is and what function a configuration management team should play within an organization or project.

I was once on a consulting assignment with a company that was going through a phase of re-designing their IT organization, bringing IT internal to the firm where it had been outsourced for a time. Management knew that a configuration management team was needed but they were not sure, or could not agree on exactly what the function would deliver. This was not a software development department, more internal IT operations, so the role of configuration management was not quite so obvious to them, given that the only function they could agree on was version control.

So they brought in a consultant, to provide an independent face that could interview all senior managers and come to some conclusions as to what everyone understood configuration management to mean, in functional terms, and what parts of that function the departments needed to establish. In this way no one was embarrassed by not knowing what to do, all reputations remained in tact.

The definitions that the consultant obtained from the managers when consolidated represented nothing coherent. At the end of the day a steel cabinet was moved into one of the offices. They put all the software CDs they had in it and designated someone to keep track of who took what, and make sure people put the CDs back when they were finished, a task which was subsequently done with little effect. This was a large organization. They were doing configuration management, to some degree; they just were not very good at it.

Configuration management can be formally defined as follows:

Configuration Management is a function that applies business, technical, and administrative direction to the development, delivery, and support of products and processes. It is the monitoring and control of both functional and physical description of a product or process throughout its life cycle.

Breaking this down, they key words are as follows:

- Direction;
- Products and processes;
- Control;
- Functional and physical;
- Life cycle.

Configuration management is primarily a logistical function and the quality of one's logistics is a good measure of how well business is being run. It is through logistical process development and control that configuration management provides direction in the business of software development solution provision, for example. Note that we are talking about business logistics here, not the freight variety.

Configuration management is concerned with the control and administration of processes as well as products and all the individual elements, or components, that make up those products. I'm not so much referring to processes that are developed for delivery to the client, although they are certainly covered also, but the processes that are put in place internally to put control around the development life cycle. In fact most of these processes are developed and run by the configuration management team.

When we refer to the "configuration," we are talking about the nature of the collection of components and processes that make up the end products. How do we describe the organization's infrastructure at any point in time while in development or production? How do we arrive at the situation where we have the information recorded to give an accurate or rather exact description? The answer is through the controls and processes put in place and managed by the configuration management function.

We apply administration to functional elements as well as the physical elements of the products. An example of a functional element would be a requirement or a change request. These functional elements clearly drive the look and feel of the physical product and are hence indelibly linked. You cannot fully control one without the other. You may record that a physical element has changed form from version 1 to 2. You may be able to compare the two versions to see, for example, what lines of code have changed. However, without the control of the associated functional elements you cannot know the exact nature of the change in terms of the overall product. You have failed to maintain overall traceability and continuity.

There are some hefty prices to pay when traceability and functional control is not effective. If you are a solution provider and you do not know the exact nature of the configuration you have delivered then you may have delivered something less that what you have been paid for or, perhaps worse, delivered some additional functions for which you have not billed the client, and cannot bill the client because you cannot quantify the function.

To quality assure your product and to limit the amount of rework required you must apply controls to the product not just at the end but also throughout the development life cycle. Many organizations focus heavily on the controls around testing and implementation while ignoring the benefits to be gained through unilateral control of the product throughout its development.

Configuration management is very much about defining and managing the development life cycle, applying controls and measures at each stage boundary. It is not about authorizing the promotion of elements, deciding what will be released or when it will be released, though it provides the infrastructure to enable all of these processes. Configuration management is about facilitation of process, to provide a logistical base for other functional groups to carry out their work efficiently and in a controlled manner.

Configuration management, like quality assurance, is a service provided to management. The goal being to keep things in order and provide quantitative measures and records of the state of the development process and the product at any point throughout its life cycle.

7.1.1 Configuration or Change?

There is often fundamental confusion as to how these terms are used. Some see configuration and change as two distinct disciplines, most use the terms interchangeably and in reference, abbreviated to CM, and leave it up to the audience to decide.

Confusion over terminology is bad news, especially where projects are concerned. As most large projects these days are assembled with staff from all over the world with differing language and jargon, the problem compounds and can lead to some costly misunderstandings. Such a misunderstanding I encountered recently was around a disaster recovery capability implementation. The meaning of the term "cold DR" was confused between client and service provider leading to the service provider committing to a cold DR for an environment of around 30 systems with a 2-hour service level agreement (SLA) or recovery time objective. As you may realize when we talk about disaster recovery in more detail, this is a tall task.

In the case of configuration management (CM), a lack of clarity around the function being discussed can lead to such problems as the CM function being placed in the wrong part of the organization or the wrong skill sets being defined for team members. The problems are essentially very much the same as they are for business continuity teams.

The situation with CM is further confused by some who muddle the two common usages of the term "change management." In the management consulting world change management refers to the management of cultural change within an organization. Mergers are good examples of where you would find a cultural change program in operation. In the context of configuration management, the change manager has a very different role.

To summarize, we have talked about what I mean when I refer to the "configuration." The configuration being the whole set of functional and physical elements that make up the end product set. To manage the configuration is to control and record changes to it, and the change request itself is controlled and recorded as part of the configuration. Hence, managing change is an integral part or subprocess to the greater task of managing the configuration.

7.1.2   Hardware Versus Software Configuration Management

You will often come across the abbreviation SCM, which stands for software configuration management. The implication is that if you have a software project you need a software configuration manager and well, the hardware and all the rest of the pieces will look after themselves, or you need a separate role or department for the network, documentation, and other areas. I do not favor the separation. It makes no sense to single out software as more important than the router configurations or the requirements, for example. All these components that go to make up the end deliverables are indelibly linked, as they are constituent parts of the end configuration.

Software is of no use without the hardware to run it on, and vice versa. The processes for controlling both are exactly the same, only the method varies slightly. Recording a change to a piece of software involves simply recording the differences in the file elements that go to make up that product through the use of a CM tool or simply keeping copies of the differing files. For hardware there is no soft element so we must create one. A configuration or modification to a configuration can be recorded as a change request, defect remediation, or a modification to a hardware specification document.

There are few cases where a hardware change would be driven by anything other than a software change, increasing storage or processing capacity being the primary exceptions. Otherwise, software drives hardware and requirements drive software. All are of course linked, interdependent, and should follow the same change or defect remediation process.

The SCM term was borne out of the traditional "configuration management is version control" mindset and is typically used by companies providing processes and tools that do just that.

7.1.3 When Do We Need Configuration Management?

I used to have a quote above my desk that said "The single most prevalent cause of IT projects coming in over budget or not at all is poor configuration management," or words to that effect. I cannot remember who said that but I think he or she was pretty spot on. The surest way to end up with poor configuration control, equally in a project or business-as-usual setting, is to introduce your configuration management processes and your configuration manager too late in the day.

In a software development setting, a lot of people are under the impression that you do not need a configuration manager until your developers come up with the first complete cut of the code. This could not be further from the truth. This reasoning normally centers on the belief that until you get to version two of something you do not have any element conflicts and hence do not have a control problem. Again, this stems from the common misconception that configuration management equates to simple version control.

While version control does not become critical until after the first integration release, it is a good idea to start it as you go on. Putting in an end-to-end configuration management process and all the tools needed to support that process takes time, quite a bit of it. CM is a management function that should ideally be in place before a single line of code is written or server commissioned.

7.1.4 Key Benefits of Configuration Management

If you understand what CM is all about then the benefits are pretty straight forward. A list of key benefits may look like this:

1. Traceability of artifacts;
2. Auditability of process;
3. Micro expenditure tracking;
4. Availability of detailed statistical information;
5. Ability to prioritize work and selectively remediate defects;
6. Ability to reconstruct an environment to any point in time;
7. Forward release planning and delivery scheduling;
8. Centralized and standardized code management functions;

9. Enhanced quality of product (fewer defects due to standards and effectiveness of defect processing);

10. Finer granularity of project control.

Good configuration management gives you information and in any business information is the key to success. Any good manager, given enough information about the state of operation, will be able to avert almost any problem that may arise and prevent many problem situations from occurring in the first place.

Where problems relate to time pressures from client or customer the information gathered from the configuration and change management processes empowers the manager with quantitative data on which to base realistic projections and situation analysis.

7.1.5 Functional Areas of Configuration Management

The key subfunctions required within a configuration management function can be summarized as follows.

Change management: controlling the progression of change requests from inception to implementation; monitoring the full life cycle of the change and crossreferencing it with the associated element versions; to facilitate the change board.

Defect tracking: providing the process and infrastructure needed to record and monitor the remediation of defects; to configure discrete builds of the product based on a set of remediated defects and deliver these builds for testing; to facilitate a system of work management for the development organization to support the allocation of work packets based on defects and change requests.

Build management: the process of coordinating build and release content based on defects and change requests to be included; to provide a central coordinated compilation process; to maintain a clean, known, controlled build environment for the compilation to take place.

Configuration repository: to maintain a repository where all master copies of every version of every element of the configuration are stored, or represented. To secure and control access to the repository; to record the configuration of element versions that make up each given build, baseline, or release.

Deployment and distribution: to provide a secure and controlled means of distributing and deploying products to the project's technical environments and to facilitate the creation of these environments from the system level; to ensure that what is deployed is a direct reflection of the build or release held in the repository and intended for the deployment; To manage a schedule of deployments for each environment and coordinate other functions to ensure maximum usability of the platforms available.

Version control: to provide a system and culture whereby all changes to all elements of the configuration result in a revision of the version of that element; to ensure that all elements that existed at any time in the overall configuration can be retrieved as a whole or reconstructed from delta information.

Document management: to extend the functions of version control and use of the repository to include documents; in using a common document and software repository we ensure traceability, and baseline integrity is maintained. Note that the promotion model for documents will differ somewhat from software elements.

Management reporting: to provide quantitative information on the effectiveness of the organization in work processing and product quality.

It is important that all these functions are implemented and available at this level of the continuity assurance maturity model. If you are not doing configuration management well you are not ready for any more advanced capabilities.

7.1.6 Definitions

Below are some formal definitions that I have put together for reference purposes.

Configuration management is a function that applies logistical processes and controls to the discrete elements of a product or process throughout the life cycle of the product.

Any complete element or element delta that contributes to, or is a constituent of, the product, is termed a *configuration item*.

Change management is the process within configuration management that facilitates the processing of requests against a product or process that result in a change to that product or process.

Release management is the process by which a given configuration is defined and promoted, trough a series of checkpoints, to a production state.

7.2 Contingency Planning

It is important, at this level of the model, to have defined and documented contingency plans in place for all critical components. Note that this is for all components, not just IT components. Remember that a critical component could be any one of the following:

- Equipment/machinery;
- IT systems;
- People;

- Place/environment;
- Information/data;
- Communication network(s).

We have identified the critical components in terms of the critical processes and products that they facilitate. We must now think about how we can build contingency into the organization by developing contingency plans for each of these components in turn.

Note that at this stage we are looking at planning based only on the technical capabilities that currently exist within the organization. The contingency plans will constitute a tactical solution for continuity issues. These should be plans that we can use immediately and that require only minimal change to the organization and its infrastructure. These plans will be updated over time as our technical capabilities increase.

Note that these plans are intended to:

1. Plan operational business-as-usual activities that minimize risk to the component;
2. Document the process that is followed in the case of an event that affects the component.

Table 7.1 looks at some possible basis for contingency plans when considering the above types of critical component.

All plans should be written under the direction of the continuity assurance team in association with those that have responsibility for the operation of the component. Where this is a critical member of staff, the plan should be written with this member of staff and his or her immediate superior, and/or backup resource. Additional provisions may be warranted for critical staff members such as company mobile telephones, laptop computers, and company cars, for example. A person's management level within the organization is not a direct representation of criticality.

All contingency plans will feed into the creation of a parent contingency plan or emergency action plan that can be used by the crisis management team (defined below) in the event of a crisis situation.

This top-level contingency plan must also be created at this level of the model and will be continuously revised. This plan will contain:

1. Instructions on crisis classification and decision-making processes;
2. Instructions on when and how to enact the individual crisis management plans;

Table 7.1
Contingency Plan Basis

Component Type	Possible Contingency Plan Basis
Equipment/machinery	Negotiate emergency stockhold with supplier;
	Stockpile critical spare parts;
	Increase maintenance checks;
	Deploy monitoring tools.
IT systems (software)	Utilize backup strategy;
	Utilize test or development systems;
	Move traffic to another node of the system, if available and no capacity problem;
	Define manual workaround (if possible);
	Utilize any resilience or recovery system currently in place.
People	Ensure that backup resource exists (secondary) and utilize;
	Ensure that detailed documentation exists for role and allocate another resource;
	Engage services of contractor with required skills;
	Outsource role or function.
Place/Environment	Utilize other locations, where they exist;
	Utilize workplace recovery site (where this capability has been implemented);
	Move to hotel or serviced office complex.
Information/Data	Utilize backup strategy;
	Revert to paper process.
Networks	Bridge problem areas with dial-up or ISDN links;
	Reroute traffic away from problem area;
	Negotiate priority agreement with network provider;
	Contract secondary provider;
	Where telephony network is impacted utilize mobile telephones (and pagers);
	If localized, move environment.

3. Communication to participants and general communication to customers;

4. Guidelines on coordination with external agencies (e.g., emergency services).

7.3   Crisis Management

It is important that your organization forms an action plan and processes to determine how you will react in a crisis situation. You will need to define for yourselves what constitutes a crisis situation; indeed you may define several levels of crisis and define what needs to be done in the event of a crisis that falls into each of these levels.

The most important part of this function is the determination of decision-making processes and communication throughout the organization of the nature of the crisis and the actions required by staff.

A crisis management team should be compiled from key resources across the organization. This team should be of adequate size to span all core areas of the organization but not be so large in number that decision making is a convoluted process. As a guide, any more than 10 people and the process can become difficult to manage. The crisis management team will be coordinated from within the continuity assurance group but, again, will be formed from staff across the wider organization.

The crisis management team will be responsible for:

- Crisis declaration;
- Action determination;
- Communication;
- Impact analysis;
- Restoration to business-as-usual post crisis.

Whereas decisions are made by the crisis management team, the continuity assurance organization will provide management support throughout the duration of the crisis.

It is important at this level of the model to determine the exact terms of reference for the crisis management team and decide whom the team will consist of, in terms of people.

The crisis management team will have as its core operating procedure the top-level contingency plan or emergency action plan previously discussed.

7.4   Design Assurance

What I call design assurance is a function and process that fits, in part, within the configuration and change management processes. It is basically represented by one or more technical design authorities or TDAs.

This function's role is to ensure that the technical environments grow in a controlled and logical way. Standards should be defined and implemented to ensure that any change or new implementation is in line with strategic thinking. The environments should not diverge in architecture as they evolve.

For example, running an IT production environment with four different flavors of Unix in operation is fundamentally bad in terms of risk and continuity. Equally, different versions of the same operating system create divergence and confusion. With these factors comes an element of risk.

The design assurance function defines and enforces a set of design principles that may be applied to all new projects or changes to existing systems. These principles may center on particular product sets or could stipulate how a new piece of software should be designed. For example, many companies do not utilize systems that make use of remote commands in their operation. This is because of possible security-related consequences in the use of such commands.

At this level of the CAM you should ensure that this function is established within your organization. Depending on your size this could be performed by the design team within the continuity assurance structure or could be a separate group within IT. Standards should also be defined and documented at this level. You may need to commission a project in order to bring your infrastructure in line with these standards to begin with. If this is required it must be initiated at this level and be completed before entering level 7 of the model. This may impact projects required at levels 5 and 6. Care should be taken around high rates of environment change occurring where projects overlap.

7.5 Backup and Tape Management

Before we can begin to look at capabilities such as disaster recovery we must look at our existing backup methods and regimes. It is true that we may eventually implement capabilities that remove a degree of importance from backups, certainly the tape variety, but poor process at this level indicates that the organization is not at a suitable level of maturity to cope with any more advanced recovery capability.

The first step is to decide what needs to be backed up and when. For this we make use of our criticality assessments and BIA. We look at each system in turn and analyze the data stored within the system. We can classify a number of different data areas within each system, these are:

1. Application binaries;
2. Application code;
3. Static data;
4. Dynamic data.

We can run into some problems here and you will find that your design assurance function may come into play more than once. Firstly, application code should not be kept on production systems. The backup source for application code is your configuration repository, which itself should be heavily attended to in terms of backups. In extreme situations code is required in production where the application does some level of real-time compilation, based on usage functions. This is generally a sign of poor application design though, and again should be looked at in terms of design assurance.

If application binaries need to be recovered this may be done from backup or directly from the configuration repository. Backups should not need to be that frequent. The most important time to backup applications is immediately before the implementation of a change. Indeed, the taking of a full backup prior to change implementation should be a requirement within your change management procedure.

Again, static data should only be changed through change request and thus can be treated in much the same way as application binaries. The change process may not be as rigorous for this type of change and changes may occur more regularly. You will have to make a judgment call on exactly how you deal with static data.

Dynamic data is data that changes through the use of the application by users or customers. This is the most important element of the system to consider in terms of backup. Frequency will be determined by the nature of the system and the nature of the data itself. The method of backup will be largely determined by the technology you have available, both in terms of backup software and the built-in capabilities of the application or database software being utilized by the system.

Note that there are some additional problems here that may require the involvement of your design assurance function. Firstly, it is not generally sensible to run multiple types of backup software in your environment. Secondly, the functionality of the backup software may require the system to be configured in a special way. For example, you may need to store dynamic data on a separate disk partition to the application and static data. This is not unusual and it is surprising the number of people that do not read the instructions when implementing backup software. In some cases if you do not partition the disks as the backup software requires you can end up with a useless or corrupted backup.

Additionally, some systems require that the database is locked or offline before a full backup is taken.

You may decide to take either an on-line backup (another disk area) or offline backup (tape). If possible, and if you have the disk space, I would recommend both. Most organizations take incremental backups each day and a full backup at the end of the week. An incremental backup stores only the file deltas, the files or the data records that have changed since the last full backup. This is faster as the backup system is not writing as much data. You can recover to the

the only checks that add real value, both in terms of you getting the best resource for the role and also in terms of protecting the security of your organization.

Some of these checks should not be done just once. It is useful to revisit some criteria on a 3-year cycle, for all staff.

7.6.2   Deployment Control

It is imperative that any deployment to your production IT environment occurs in a secure manner. For this we can make use of the change and configuration management procedures previously talked about in this chapter. Adding an element of security to these processes will also ensure that the processes are not bypassed. Remember that error in change implementation is the single biggest cause of system failure. A change implemented without the knowledge of management can also have a devastating impact on your recovery capabilities.

The way I recommend to do this is to utilize the server or server cluster that houses your configuration repository as a central access point to any production environment. For example, configure your production servers so that they do not allow direct connection from any other terminal than the configuration server. Establish a trust relationship between the configuration server and all production servers. In this way all users have to log onto this configuration server before they can get to any of the production systems. Activity can be logged centrally on the configuration server, in the case of any deployment or configuration change. Any deployment should come directly from the configuration repository. In this way we ensure that the repository has been updated and that the code has been through the correct levels of testing before being deployed. We are assured that the image in the repository is in synch with the actual image in the production environment. This of course has added benefits in terms of recovery. Incidentally, this same system should also control all other environments such as testing and recovery, with the exception, to some degree, of development environments. The system should still control the development environment but not in quite the same way. I will not go into this any more as this is not a book on configuration management.

7.6.3   User Access Control and Monitoring

You should have detailed processes and control mechanisms around user access control. These processes should include a degree of active monitoring.

Those in IT departments are themselves often best at ignoring or working around security, which is all the more reason to separate IT security from the IT department, as I have suggested. The IT department should have a measure of control over its development systems to ensure that they can do their work. This needs to be done in a pragmatic way and in a way that is still controlled. Desktop security and production control processes should be vigorously enforced.

To maintain true auditability and integrity of process initial passwords should be sent to the individual via internal mail in sealed envelopes. Subsequent passwords and system access details may then be sent via e-mail. The line managers should not be directly involved in this process, other than to make the initial request.

Usage should be monitored to some degree and unauthorized access attempts to any system should be logged and followed up by a member of the security and safety team.

Pay particular attention to information leaving the premises, either via the Internet or on laptop computers or mobile disk drives. Information should be categorized and some should not be allowed to leave the premises at all. While enforcing these types of rules again ensure that you remain pragmatic and do not discourage the use of new technologies or inhibit anyone's ability to do his or her work. Finding the point of balance is what continuity assurance is all about. If your security is too lax, you will have issues with continuity. If it is too tight, you will also cause problems that may affect the continuity of your business.

7.6.4   Data Center Security

Your data center should be one of the most secure locations within your organization. A data center should be physically isolated from any other office space or working area.

Where possible, a data center should be purpose built and take into account the following:

- There should be no windows in the data center.
- The building should be of solid construction.
- Power and network connections should be redundant to the front and back of the building.
- Environmental controls such as air conditioning, air filtration, humidifiers, or dehumidifiers should be fitted. Note that the requirements here will vary as you add or remove systems on the data center floor.
- The building should be fitted with adequate uninterruptible power supply (UPS) equipment. This should be physically separated from the server rooms.
- All server rooms should have raised floors with cabling under the floor.
- The building should be equipped with a fire suppression system utilizing an inert gas or similar. Standard sprinklers are clearly a bad idea. This system should also be physically separated from other areas.
- At least one backup generator should be available with adequate fuel to run for a reasonable time period, usually 2 to 3 days. In the event of

an extended outage priority contracts should be in place with fuel suppliers.

- The building should be designed with a loading bay incorporated for the delivery of emergency equipment.
- An on-site tape storage facility should exist within the building.
- A store room should exist within the building for the storage of critical spare parts.
- Water tanks should not be situated on the roof or in the ceiling space.
- Ablution facilities should be situated on the ground floor.

Consider the following points when assessing or implementing security within your data center:

- Your data center should have a single entry point for general use (apart from emergency exits and loading) and this entrance should be controlled by special security pass and a physical presence (guard).
- Visitors should be authorized ahead of time and, depending on the nature of your facility, be security checked. They should be escorted at all times.
- No equipment including disks, laptops, or mobile drives should enter or exit the data center without written authority.
- Consider banning the use of mobile telephones within the data center, depending on the nature of your equipment.
- The data center should be equipped with recorded CCTV running 24/7.
- All doors should be of solid construction and self–closing.
- Emergency exits should be constructed of reenforced steel.
- All equipment should be racked and cabinet doors closed and alarmed.
- Implement a pest-control regime for the building; mice and other rodents have a habit of chewing on cables (I know of a feral cat found in one major data center).
- In general, vehicles should not be allowed to approach the facility under general operating conditions.

7.7 Level 4 Performance Indicators

Table 7.2 sets out the performance indicators at this level of the CAM.

Table 7.2
Level 4 Performance Indicators

Rigor	
Item Number	**Action/Task**
R4-001	Implement or upgrade change and configuration management processes and infrastructure to the standard outlined within this methodology. Migrate all information to the configuration repository.
R4-002	Define and document contingency plans for all critical components.
R4-003	Compile an emergency action plan.
R4-004	Form a crisis management team and document terms of reference.
R4-005	Establish a design assurance function and associated design principles, processes, and standards.
R4-006	Initiate the transformation of infrastructure such that it is in line with defined design standards. This should be completed before entering level 6.
R4-007	Review backup and tape management procedures. Determine critical data. Ensure that a defined backup regime exists and is followed. Backup and restore should be tested and contracts should be in place for secure tape transport.

Management Functions	
Item Number	**Action/Task**
M-016	Ensure process changes are adequately communicated throughout the organization.
M-017	Ensure that documentation is compiled in an auditable fashion and added to the configuration repository.
M-018	Continue governance of projects commissioned at earlier levels.

Security and Safety	
Item Number	**Action/Task**
S1-009	Review all security procedures for rigor of process.
S1-010	Define and implement resource vetting processes.
S1-011	Assess deployment management processes in terms of security and remediate issues.
S1-012	Review user access control procedures and processes and remediate any security issues. Ensure processes include active monitoring.
S1-013	Define data center security infrastructure and processes and implement them.
S2-007	Review all safety procedures for rigor of process.
S2-008	Define data center safety infrastructure and processes and implement them.

8

Level 5: Robustness

This level of the continuity assurance maturity model is concerned with robustness of architecture. It is primarily related to IT systems but could be equally applied to building designs, site layouts, or other equipment used in the delivery of your business functions. Our discussions here will focus on IT systems.

Before we implement specific capabilities to assure continuity, such as resilience and recovery capabilities, it is necessary for your organization to be at a certain level of maturity to best leverage these capabilities, or indeed to make a success of their implementation at all.

If you attempt to recover an architecture that is diverse and not coherent in its design you will have a big job ahead of you and you will almost certainly spend more money than you would have to if you first addressed your overall architecture. If your architecture is solid in its foundation then you may not have an event that impacts you enough to have to initiate recovery. Indeed, a simplified, rationalized, or streamlined architecture means that some continuity impacting events may not occur to begin with.

We will discuss a number of initiatives that will raise the level of maturity of your organizations IT infrastructure and make the implementation of resilience and recovery capabilities much simpler. These initiatives should also decrease the overall time to implement such capabilities and increase the probability of success of both implementation, and of use, in the event of a continuity-impacting event.

Specifically, in addition to security and safety considerations, we will discuss:

1. Server consolidation;

2. Storage area networks;

3. Enterprise application integration;

4. Common operating environments;

5. Distributed systems;

6. Infrastructure renewal.

Note that we are not talking about resilience at this level of the model. Resilience is the subject of the next chapter. These initiatives focus on preparing the landscape for the implementation of resilience and recovery capabilities.

Under no circumstances would I recommend proceeding with any of these initiatives at the same time as implementing advanced recovery capabilities. There are reasons why these tasks must be done in sequence, the least of which is the fact that this would be an extremely high-risk endeavor in itself. Rationalize your infrastructure, let it stabilize, and then implement more advanced recovery capabilities.

8.1 Server Consolidation

There are many arguments to be put forward both for and against server consolidation. Server consolidation involves moving applications from individual, normally smaller servers often dedicated to that application, onto larger servers where multiple applications can be hosted in parallel. The applications are configured onto what I will call workloads, which are discrete areas of disk within the consolidation server. The consolidation server is configured in such a way that each workload, virtually speaking, behaves exactly the same as a separate server would. In theory, there is no impact to the application because, again in virtual terms, it is residing on what is for all intents and purposes, its own server.

As with all things it is not quite as simple as all that and there are applications that have problems with some consolidation platforms. It all depends on the architecture of the consolidation system and how compatible this is with the way that your application has been designed.

If you are going to go down the consolidation path make sure you use hardware that is specifically designed for this purpose. Proper consolidation does not mean taking two standard servers, each running their own application, and putting both applications on one of the servers. Many people attempt this of course and run into all sorts of ugly problems eventually. Apart from functional problems the capacity planning and security issues do not bear thinking about. Clearly associated applications that are designed to work together are a different story, as are agents and the like. Here I am talking about major applications.

Server consolidation can be a tricky exercise but the benefits are self-evident. If you consolidate 10 small servers onto one big server then you only need to have another big server in your recovery site, in simple terms. There is also, of course, a lower cost of ownership of one server as opposed to 10. Additionally, the overall footprint will normally be smaller so it takes up less space in your data center. Be aware that the lower cost of ownership means that the cost of purchase is high, in some cases very high indeed.

There are some problems though. You will be introducing a single point of failure into all of the systems you consolidate on each consolidation server, the single point being the server itself. This may be acceptable but if you have resilient servers for the systems that you are consolidating do not even think about consolidating them on the same unit as the primary node.

Consolidation systems can be proprietary in terms of operating system, so if you have a small operation with high diversity in terms of the number of operating systems you are running then consolidation is probably not for you.

8.2 Storage Area Network (SAN)

Storage area networks are a means of storage consolidation. All systems that store dynamic data will obviously have some type of storage for that data, whether it is on the local disks of the application or database server, or by some means of direct attached storage (DAS). It is useful to collect all this data and store it in one place. This place is the storage area network.

There are huge advantages to be had from the use of a SAN, in terms of continuity assurance. Briefly, these are.

- Ease of replication if the SAN infrastructure bridges multiple sites, that is, high availability or recovery environments. SAN technology uses block level replication which is often much simpler and easier to manage than application level replication.

- Ease of backup to disk or tape library directly attached to the SAN infrastructure. In this way the SAN can facilitate a centralized backup capability.

- The SAN is more straightforward to manage in terms of capacity planning.

The implementation of a SAN at this level in the continuity assurance model will transform your IT infrastructure landscape in a way that will be invaluable in terms of implementing resilience and recovery at later stages. It will vastly improve your current baseline recovery capabilities.

You do need to be aware that a SAN implementation and migration project is far from a trivial exercise, and it will be very costly. Not only do you need to purchase and implement the technology but also you will need to assess each of the applications that you wish to migrate and put a plan in place to move the data. In some cases you will need to upgrade your server or operating system to ensure compatibility and supportability with the SAN technology. This migration task is simpler where direct attached storage is already being used but there are risks.

To ensure that the application is not impacted in terms of performance you will need to connect the servers to the SAN via a fiber channel network. This will require a fiber channel card, or two, in each server and some fiber channel network infrastructure (e.g., switches). Fiber channel is much faster than ordinary Ethernet.

If you are running particularly old hardware you may be forced to upgrade before you implement a SAN. We will discuss infrastructure renewal a little later in this chapter.

It is possible to migrate entire systems to the SAN. This means that the operating system, application, and data are all moved to the SAN. Unlike server consolidation the actual server that you are migrating from remains, it just becomes diskless. The server operates as usual in terms of processing; it sees the disks on the SAN as if they were local to the unit. This type of migration is supported by many SAN providers but I would not recommend it. In my opinion it is best to keep at least the operating system resident on the individual servers, everything else can be migrated to the SAN.

Note that only some versions of operating systems can be migrated to the SAN. In general terms, another advantage of the SAN is that it can deal with all flavors of operating systems, unless you are running something particularly exotic.

The disadvantages of utilizing SAN technology can be summarized as follows:

- It is expensive;
- You may be left with a lot of redundant equipment, like DAS disk packs;
- You are introducing another single point of failure (SPOF).

8.3   Enterprise Application Integration (EAI)

EAI is a pretty general term. It basically refers to the harmonization of all of your systems into a single integrated megasystem. It is focused on improving and standardizing the methods of interface between your systems so that they all act in harmony as one integrated business enabler. To achieve this you will need to make, often major, modifications to every system you have, at the application level.

EAI is an idealistic approach and EAI projects in general have a high failure rate. I do believe that it is the way to go though and implore you to have a go at it. The most important thing is to understand what you are getting into. Understand and have respect for the complexities of such a project and you will be okay.

There are many systems that you will utilize in your business and most of them will be created by a range of different companies. Realize that they are not all designed in the same way and they are not all necessarily designed to talk to each other. This is largely why we have a relative plethora of companies out there calling themselves system integrators. Most companies will implement a system and develop their own specific adaptors on a system-by-system basis to enable all the systems to talk to each other.

EAI is generally thought of as the next level of evolution beyond this in that with EAI we are implementing a common messaging layer that operates across all applications within your landscape. There are a number of companies that produce software to enable this.

If you are thinking of setting off on an EAI project then it needs to be done at this stage of the model, before you implement more advanced recovery capabilities. EAI will alter your system's infrastructure in a fundamental way. Making these changes across multiple environments is problematic and costly. You may also find that you would have approached your disaster recovery environment differently if you had EAI in place before you began.

The key benefits of EAI, in continuity assurance terms, can be summarized as follows:

- Overall, produces a more scalable infrastructure and simplifies the integration of new systems;
- Provides a level of built-in resilience in your application messaging layer;
- Gives the ability to integrate across multiple sites/environments;
- Provides more efficient transaction processing and enables transaction recovery;
- Can be implemented in a way that prevents the occurrence of in-flight transactions at the time of disaster.

The points against it are:

- High cost;

- Can take a long time to implement;
- Large change to infrastructure and hence high risk.

8.4   Common Operating Environment (COE)

This is by no means a new concept. Most large Western organizations implemented common operating environments throughout the 1990s but there are still some organizations out there that have not embraced this yet, or at least have not implemented it as well as others.

A common operating environment is most often thought of as a common user desktop, basically a standard workstation built with the same operating system version and a standardized suite of applications available. This is particularly advantageous in terms of ease of support. Having a standardized build also adds a degree of overall stability to your IT infrastructure. Security around the change of the desktop means, for example, that you do not have people loading the latest and greatest version of Windows on their computer workstation, perhaps causing incompatibility issues with other company software or hardware.

A COE also has its advantages where continuity assurance is concerned. Apart from the stability of infrastructure and hence reduced risk of error or incompatibility, there is also an inherent advantage in all systems being the same. This advantage relates to the ability of a user to move to another location and use another person's terminal because they are familiar with the desktop and the tools. Depending on how your network has been set up, users may also have access to their own personal profile over the network. This means that they can log onto any workstation in the company and have access to their own personal settings, mail, and files. This has clear benefits where workplace recovery is concerned.

The next generation in common operating environments comes with terminal server technology such as Citrix. Citrix is basically thin client technology for Microsoft Windows. It allows users to connect to a Windows desktop via a diskless terminal. All file storage including operating system, desktop applications, your personal profile, and your working files is resident on the Citrix servers. Everything on these servers is backed up regularly, according to your backup regime, and you can access your work from any terminal in the organization in exactly the same way you would at your own desk. This sort of system is absolutely brilliant in terms of continuity assurance. As long as the server infrastructure can be recovered all you need is a room full of relatively inexpensive terminals to get working again. There are a number of other benefits to the organization in terms of security, supportability, capacity planning, scalability, and so on.

The disadvantages of this sort of system and to a degree any type of COE are as follows:

- Depending on the distance between server and client, and on the characteristics of your network, there may be performance issues.

- You have, to some degree, introduced a SPOF. If a Citrix server goes down all the users connected to that server will lose access to their desktop.

- Some specialist applications require a great deal of configuration or customization to get them to work in this sort of environment.

- Users can become frustrated with the use of superseded software versions if your IT department is a little slow on the upgrade path. Indeed, users may be prohibited from using new technology that would better the interests of the company because of compatibility or "time-to-implement" issues.

We have talked a lot about the desktop environment but COE is not just restricted to desktop infrastructure. General server infrastructure can also be standardized in terms of server build. This is particularly effective for Wintel–based servers but can be equally applied to all. These sorts of build standards, however, should have been addressed under design assurance, which was discussed at the last level of the continuity assurance model.

8.5 Distributed Systems

Distributed processing can be good for continuity. A distributed system is a system with multiple servers or nodes. Each node runs the same application but services a different customer or user range. The range can be set by any number of factors including geographical location of customer or user, or can be completely arbitrary. It may simply be the case that the next user connecting is routed to the next available node with spare load capacity.

There are three ways in which distributed systems can assist in maintaining continuity. These are, in no particular order:

- If one node fails the others will continue to operate. Only the customers or users of that particular node will be impacted.

- If a node fails it may be possible to divert traffic from that node to a functional node or nodes. Due to the added load the remaining nodes may be impacted in terms of performance but this may be more acceptable to the business than a complete stoppage of service to a given customer or user group. You may design a degree of emergency capacity in the overall system so that the system, across all nodes, can cope with the failure of at least one node at any one time.

- By adding a degree of geographical diversity to the nodes, by situating each node in a different location, you have removed a SPOF from this particular system, the SPOF being the common location. You have therefore reduced risk to the overall system.

It is important to work out which systems you will distribute at this level of the model. You will need to have this information to be able to design your resilience capability for critical systems. Distribution is a way of building a degree of resilience into the design of a system.

Note that geographical diversity may mean putting each node in a different data center. If you are a global organization it may mean putting each node in a different country. The bigger the distance between nodes the more problematic and costly this can become, but the bigger the benefit in terms of continuity. You are unlikely to suffer an event that hits you simultaneously across three continents. It is possible though; software viruses are a good example of where location will be largely irrelevant.

8.6 Renewal

Prior to the implementation of any resilience or recovery infrastructure it is important that you take a hard look at the age of your production infrastructure. If your infrastructure is at or near end-of-life in terms of supportability you may need to undertake a significant infrastructure renewal project before you are at a sufficient level of maturity to move forward with continuity assurance.

Old hardware in your production environment can be a problem for a multitude of reasons:

- If the hardware becomes unsupportable, that is, the manufacturer says that they will no longer support it, you will be faced with having to support it yourself or contracting a third party to support it. Neither is likely to be cheap.

- As the hardware becomes older the frequency of failure may become more frequent or more likely, which is of course bad for continuity.

- You may need to upgrade an operating system, for example, because the newer version is required to be able to migrate to a SAN. This new version of the operating system may not be compatible with your old hardware, so you will be forced to upgrade.

- If you are purchasing a new system for resilience or recovery, you will not be able to purchase the same model new as that which you currently have, as the manufacturer will no longer be making it. This means that you are faced with one of two options:

1. You could have a mirror system that is running a different OS to your primary, which may in turn mean a different version of your application. The higher the degree of difference between primary and secondary system the higher the likelihood that it will not work when you need it to.

2. You could purchase second-hand hardware that is equivalent to your primary production equipment.

The second numbered point above is always a tricky one. If you are in this situation you need to make a decision based on risk. If you purchase the second-hand hardware you could have yourself a working resilience or recovery capability very quickly and without a lot of change to your production system. The downside is that the cost of ownership will be high and there is a risk that the equipment will fail. If you go down the renewal path and upgrade the production system, and implement an exact replica for resilience or recovery, your overall continuity will be stronger, as long as you do not have any problem with the production upgrade.

Renewal is often necessary and is pretty much a fact of life. The problem is that it is in itself a high-risk exercise. Given that upgrade is usually inevitable at some point the best way to minimize this risk is to go through this renewal exercise as regularly as is practicable. While it is not sensible to implement every new version or model that comes along it is practical to follow one to two models behind the curve. As we are principally concerned with the effect of upgrade on our applications, your infrastructure renewals should be paced against operating system versions rather than hardware models. New operating system versions do not come along every week and keeping one to two versions behind should give you a 3 to 5-year gap between hardware renewals, unless of course you are driven by the implementation of new application software to enable your business. Over 3 to 5 years you should have fully depreciated your hardware anyway, depending on which country your business is based in.

8.7 Security and Safety

In looking at the robustness of infrastructure, or in other words the ability for the infrastructures overall design to resist attack and remain intact, security plays a major role.

We have talked about standardizing the infrastructure in general design terms and there are some inherent security factors that need to be part of these standards. The most important in continuity terms is what is called "server hardening."

Server hardening involves configuring the server such that connection to it, and communication from it, is restricted to that which is essential for the functionality of the application. Server hardening will include altering the server's access control list, closing communication ports that are not used, and modifying file permissions. A standard should be defined so that certain high-risk configurations are remediated for all systems in the same way. In some cases the application may need to be altered so that the standard can be met.

High-risk configurations may be those that, for example, leave file transfer protocol (FTP) ports open or have a multitude of users registered directly on the server's access control list (ACL). For UNIX–based systems, being able to logon from another system as the root user is generally thought of as unacceptable, for example.

We have talked about the use of the configuration management system as the central access point for all production systems. If we do this we do not necessarily need to have individual users on each of the systems. The users for each system may be limited to an administration user (or root), an operations or monitoring user, and a deployment user, in addition to any specific users required by the applications. The ideal is that individual users are logged at the configuration management server. We know who is using the deployment user on a given system because we can see who has actioned the logon command from the configuration management server to the particular application server. As previously mentioned you should not be able to connect between application servers directly.

Communication between the configuration management server and application servers should be by secure shell (SSH) or similar secure communications protocol. SSH is essentially a more secure version of Telnet. It means that you do not have to have Telnet active on the application servers. Note that although SSH is generally thought of as being shareware the version for commercial use is not. You will have to purchase licenses for every server.

As a general principle, shareware software should not be used at any level within your organization. This is for reasons of security and liability as the code will often be open source. Additionally, shareware is often not readily supportable, that is, obtaining support agreements can be difficult or in some cases impossible.

In terms of safety, at this level of the continuity assurance maturity model you should ensure that you have your safety infrastructure, or in other words your equipment, procured and implemented.

We have talked about such things as provisioning your own internal emergency services, where warranted. If you are going to do this then it should be completed at this level.

You should assess the number and location of fire-fighting equipment, extinguishers, and the like, within your buildings. Do not just look to see if you have met the regulation. Think about the most likely places for fire to develop and how you would go about defending against the fire. Place equipment in locations where you think it will be most useful. For example, do not put fire extinguishers behind doors where they may not be clearly visible. I am sure your local fire service will be only too happy to help with this exercise. A formal fire-risk assessment should be carried out.

First aid kits should be available also throughout the building in logical and visible locations. Clearly, these should be fully equipped and checked to ensure all components are within their usage dates. Oxygen is also useful to have on hand.

Depending on your risk profile, consider equipping your offices with a supply of gas masks, flack jackets, hard hats, or chemical suits. This may sound extreme but this sort of equipment is almost impossible to get hold of when you need it. I recall the Anthrax scare in 2001 where organizations and families alike were seeking out supplies of gas masks and chemical suits in case of biological attack. I was in Australia at the time and army surplus gas masks were selling for up to $300 U.S. dollars each. These sorts of threats are real and represent a danger to us all. Better to face up to this and be prepared rather than be ignorant and be dead.

I will mention, as I have before, that it is possible to fit your workplace with chemical sensors that operate in a similar way to smoke detectors. If you feel that you are at risk then this is the time to get these implemented.

Remember that if you cater for members of the public at your business premises then you have a moral responsibility for their protection against any identified risk while they are in your care or on your premises. I would recommend that you cater for them in the same way that you cater for your own staff. This is of particular relevance to organizations such as train operators, for example. Trains, particularly the underground variety, can be a deathtrap in the event of terrorist strike. This is not fiction; it has happened and is thus for many operators a known risk. Be responsible, be alert, be safe, and stay alive.

8.8 Level 5 Performance Indicators

Table 8.1 sets out the performance indicators at this level of the CAM.

Table 8.1

Level 5 Performance Indicators

Robustness	
Item Number	**Action/Task**
R5-001	Conduct an audit and a design review of your overall infrastructure.
R5-002	Assess the costs, benefits, and feasibility of server consolidation and commission projects to implement were applicable. (A maximum rating of 2 can be given in the case that this capability is not implemented, see Chapter 11 for rating details.)
R5-003	Assess the costs, benefits, and feasibility of storage consolidation and the implementation of SAN technology. Commission and complete projects to implement were applicable. (A maximum rating of 2 can be given in the case that this capability is not implemented, see Chapter 11 for rating details.)
R5-004	Assess the costs, benefits, and feasibility of implementing EAI enablers. Commission and complete projects to implement were applicable. (A maximum rating of 2 can be given in the case that this capability is not implemented, see Chapter 11 for rating details.)
R5-005	Assess the costs, benefits, and feasibility of implementing or upgrading COE and terminal server technology. Commission and complete projects to implement were applicable. (A maximum rating of 2 can be given in the case that this capability is not implemented, see Chapter 11 for rating details.)
R5-006	Assess the costs, benefits, and feasibility of distributing systems and commission and complete projects to implement were applicable. (A maximum rating of 2 can be given in the case that this capability is not implemented, see Chapter 11 for rating details.)
R5-007	Conduct any infrastructure renewal projects that may be required and put in place an ongoing renewal strategy.
R5-008	Complete harmonization of technical environments with design standards, initiated at level 4. Take account of any changes arising from developments at this level of the model.

Management Functions	
Item Number	**Action/Task**
M-019	Ensure that all changes are managed through the change control processes and are reflected in the configuration repository.
M-020	Ensure changes are adequately communicated throughout the organization.
M-021	Continue governance of projects commissioned at earlier levels.
M-022	Conduct a quality review of any project completed while at this level of the model.

Table 8.1 (continued)

Robustness	
Security and Safety	
Item Number	**Action/Task**
S1-014	Define policies and configurations for, and implement, hardening of all servers.
S1-015	Implement a secure communications protocol between servers, such as SSH.
S1-016	Remove any shareware software used within your infrastructure. Replace with licensed software or redesign to remove the need.
S2-009	Plan, procure, and implement safety equipment/infrastructure.
S2-010	Determine the most sensible location for each piece of equipment and install.
S2 011	Review or implement processes for the maintenance, restocking, and testing of safety equipment.

9

Level 6: Resilience

Resilience is defined as the ability to recover quickly from unpleasant or damaging events. Hence resilience is a form of recovery, but the key word in this definition is "quickly." It is generally accepted in IT circles that resilience refers to mirror or fail-over systems. Whereas recovery systems typically will be located in a separate facility to the primary system, a resilient system may be co-located with the primary system. There are a number of exceptions to this, which we will discuss further in this chapter and the next.

The key difference between resilient systems and general recovery systems is that the resilient system is part of the production infrastructure and is permanently on-line and up to date in terms of data, or at least as up to date as it needs to be to function within an acceptable tolerance. A recovery system will generally need some tasks performed to prepare the system and redirect traffic to the system. The extent of work required on the recovery system will depend on the particular capability configuration, which we will discuss shortly.

Resilience is implemented more to guarantee a level of system availability to customers and users, in terms of system related problems and the performance of maintenance. It is not usually used as a means of recovery from disaster, a disaster not generally being a localized event, local to a single system that is.

It should be noted that in the context of this book we consider recovery to be the capability of recovery from a disaster event. Recovery infrastructure may be utilized in non-disaster events but the capability is designed to meet the requirements of just such an event. For this reason recovery will always be enacted at an alternate location.

9.1   Capability Configuration

It is necessary at this point to go through the different levels, as I see them, of capability that we may implement. These capabilities can be equally applied to resilience and recovery.

In order of best-to-least effective in continuity assurance terms the capability configurations are as follows:

1. Reciprocating load;
2. Reciprocating data;
3. Hot standby;
4. Warm standby;
5. Cold standby.

Table 9.1 defines each of these levels of capability.

In Table 9.1 I refer to a "heartbeat" between systems. This is a mechanism inherent in cluster technology. The secondary system continuously poles the primary system to see if the system is running. If the system stops functioning the secondary system will take over. The same applies if there are more than two nodes. This technology can be configured in a nested or interlaced fashion, that

Table 9.1

Capability Configurations

Capability	Defining Characteristics
Reciprocating load	A clustered system where the load is balanced between a number of nodes. Fail-over should a node fail is automatic and all nodes are capable of taking the full load. Data is stored in a common system which has a reciprocating data standby capability.
Reciprocating data	A system where both hosts are on-line and clustered but only one is active. Transactional processing is done in a synchronous fashion between the two hosts so that the second host is always current. There is a heartbeat between the systems that switches traffic to the secondary system automatically if the active system fails.
Hot standby	The alternate system in on-line and clustered, there is a heartbeat between it and the primary system so fail-over is automatic. Data, however, is replicated to the secondary system in an asynchronous fashion.
Warm standby	An alternate system exists, is on-line and is up to date in application software terms but data is not current. Fail-over is a manual process.
Cold standby	An alternate system exists but requires boot or recovery from media. Fail-over is manual.

is, fail-over to one then the next and so on, or fail-over to all remaining nodes in parallel with the load split between the remaining nodes.

Capabilities are recommended based on the criticality rating of the system concerned. Table 9.2 shows the recommended capability for each of the levels of criticality.

Note that for ancillary systems a recovery capability need only be put in place where there is a primary or secondary system that is dependent on the ancillary system.

I have illustrated in Table 9.2 the capabilities required for the different levels of a critical system. Do not be confused into thinking that if you have a resilience capability you do not need a separate recovery capability. In most cases you will need a lower-level capability for recovery than resilience, because of longer recovery time objectives allowing you more time to configure your solution. We will go through this table again when we talk about recovery. For the moment we are concerned with implementing resilience capabilities.

9.2 Replication

Before we can move on you must first understand the difference between asynchronous and synchronous replication.

Many people believe that asynchronous replication only occurs when you have some sort of batch transfer process that only transmits changes to a

Table 9.2
Recommended Capabilities

Capability/Criticality	Primary	Secondary	Ancillary
Reciprocating load	Where no outage is acceptable.	N/A	N/A
Reciprocating data	Where data integrity is an issue.	Where data integrity is an issue.	N/A
Hot standby	Minimum capability for resilience.	Recommended.	N/A
Warm standby	Minimum capability for recovery.	Minimum Capability for resilience.	Recommended.
Cold standby	N/A	Minimum capability for recovery.	Minimum Capability for resilience, and recovery (only where system is dependent).

secondary system intermittently, say every hour, for example. Yes, batch transfers will always be asynchronous but this does not imply that all "real time" transfers are synchronous.

Consider two databases that you wish to keep in step with each other as part of your resilience capabilities. As a transaction is committed to the first database the transaction is immediately sent to the second database and committed. However, there is a time lag between the transaction committing on the first database and it committing on the second database. This time lag is due to network latency and processing time. This is asynchronous replication. What would happen if the primary system failed during the time in-between committing on the first database and committing on the second database? When the traffic was moved to the second system the database would not be current to the point of failure.

In synchronous replication the transaction is queued on the first database and sent to the second database. Once the transaction has committed on the second database an acknowledgement is sent back to the primary system and the transaction commits on the first database. In this way we know that any transaction on the first database has already been processed by the second database. There are of course minor variations in the way in which this processing occurs depending on the technology involved but, for the sake of example, this is synchronous replication.

Replication is of course not limited to databases and there are a number of software and hardware-based technologies that exist to facilitate it. Most databases have their own replication mechanisms but this does not mean that they cannot be replicated using other technologies also. Many packages replicate at low levels, for example, block information on the systems disks.

9.3   Data Integrity

It is important to understand the issues that you may face in terms of data integrity. Data integrity issues are inherent in asynchronous replication; indeed, this is always a major part of the justification for using synchronous technology. However, data integrity can still be a problem even where synchronous replication is being used.

Firstly, if we use asynchronous replication we will have the situation where the data on the resilient system may not be current to the point of failure. If we are in the situation where we have multiple related systems, that is, systems that interface to each other, then it is possible to have data inconsistencies in-between these systems caused as a result of the interruption of the interface. For this reason, when we recover multiple systems we will usually roll back transactions on all systems to a point where we know that the data is consistent across all systems. This may result in a larger data loss or longer recovery point than we may have hoped for. This has a particularly large impact if we have some batch-based interfaces that only run once a day. You may find that you have to roll back all connected systems to a point before this process last ran.

Even when we are using synchronous replication we can encounter problems, especially if the applications have not been designed to account for the particularities of working in a "high-availability" situation. Consider the example in Figure 9.1.

Let's say that a purchase transaction is initiated from a Web site. The user has already keyed in his personal data and details of the items he wishes to purchase, and hit the confirm key. A transaction is initiated that has to interact with a number of different databases or systems. These could be separate databases within a single system or separate systems all together. If there is a failure that occurs somewhere in-between system two and system three then we have a problem, even if all three systems utilize synchronous replication for their secondary systems. We could be in the situation where the stock has been picked and sent to the customer but the customer has not been billed. If we roll back to a consistent point across all three databases then we may lose the evidence of the transaction ever taking place. One would hope that a roll back would highlight the inconsistency and we could trace the problem and bill the customer but things are not always this easy and systems are not often as simple as the example above. This type of occurrence is called an in-flight transaction.

This situation would not occur if the software architecture took into account possible failure within its design. The way that this should work is that

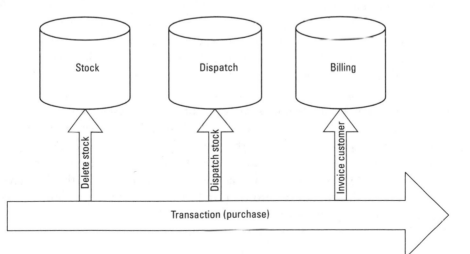

Figure 9.1 In-flight transactions.

the transactions are queued on each of the databases and once an acknowledgement has been received the transactions should all simultaneously commit at some designated time in the future. They are set to confirm in the future to account for any delays in processing. This may be 1 or 2 seconds into the future.

Use of a middleware product such as those often implemented as part of EAI will solve this problem in most cases, or at least help the situation by enabling a higher degree of control around transaction processing.

Be aware that when we are talking about replication time is very important. Every network has a service running somewhere on the network that is a time-server. The purpose of this time-server is to make sure that all the systems on the network have exactly the same time. If time is out of synch between interfacing systems you will get inherent data inconsistencies. A simple example of this is where you receive e-mail messages that appear to have been received before they were sent. Imagine the problems this sort of thing can cause with financial management systems.

9.4   Data Redundancy

It is necessary that you have a high degree of understanding of your overall data architecture so that you can determine what data needs to be replicated and what is superfluous. There are other considerations here and you may find that you need to replicate redundant data just because of the way the systems are integrated or the time to regenerate data from first principles takes too long to meet time or point objectives.

By far the lowest risk strategy, particularly if your understanding of the data architecture is poor, is to replicate everything. This is of course not the most efficient approach and demonstrates a lack of understanding that in itself could be dangerous and pose a future threat to continuity.

I have encountered many organizations where the same data is passed from system to system and undergoes some transformation at each step in-between. In this sort of scenario you should look at the first and last system in the chain. The last system will probably have the bulk of historic data. You need to ask yourself whether this data needs to be available on-line or whether it can be archived to tape or SAN and retrieved as and when required. Does the data have a role to play in day-to-day processing? If it does not then it will probably not be required on-line. You will need to assess the time lag between the data moving from the first to the last system and see if this fits inside your recovery-point objectives.

You overall data architecture should be revisited with a view to optimizing data storage. Rather than trying to pick what is and is not required in a resilient or recovery system, you should try to store on-line in production only the data that is required. This data should be stored in one place only in the production systems. Yes, having the same data stored in multiple systems could be an asset in continuity terms but this is why we have resilient and recovery systems. The data will be in multiple places, but it will also be accessible and well understood.

In summary, it is essential to rationalize your data architecture before embarking on the implementation of resilience or recovery capabilities. Once this is done you can take the lowest risk approach to replicating all stored data, perhaps with the exception of historic, nonessential, data.

9.5   Geographic Diversity

In terms of resilience, there are two ways that we can add a degree of geographic diversity. Geographic diversity means that the resilient system is not in the same physical location as the primary node. In most data centers you will find the resilient system next to or in the same cabinet as the primary. This means that if there is a problem that affects the data center as a whole, or is localized to that particular section or cabinet, then we have lost, in addition to the primary system, our resilience capability for this system.

Adding a measure of geographic diversity may simply mean placing the resilient systems on the other side of the data center or in a high-availability server room within the data center. High availability is a term that is often used interchangeably with resilience. Some organizations will split the data center into two distinct sections with a physical fireproof wall running between them. On one side of the wall are the production systems and on the other are the resilient or high-availability systems. This is a particularly cost-effective way of obtaining some high-value geographic diversity.

Of course the optimum solution is to configure all resilient systems into a separate high-availability data center. This will probably need to be reasonably close to the production facility due to technology limitations concerning clustering. The production and high-availability sites will need to be linked with a high-bandwidth dark-fiber network. This is expensive, justification for this should be based on the resilience capability configurations required and the risk profile of the production data center. It is, however, the optimum solution in terms of continuity assurance. Note that this high-availability site does not necessarily do away with the need for a separate recovery site. If your requirements are such that you can get the production and high-availability sites far enough apart you may be able to utilize this as a dual resilience and recovery site, but this is unlikely to occur when working under the requirements of this methodology.

We have previously talked about distributed systems. Distributed systems are a sound way of effecting geographic diversity. Again the possible distances involved can be problematic when dealing with the replication of data. The best candidates for distributed processing are systems where no or little data is stored, that is, systems that are primarily used for processing only, and where any data is passed to downstream systems. If you wish to utilize the geographic diversity of a system that is already distributed within your organization you may have to do some work to configure the fail over sequence between nodes. The fail over may need to be manual, again because of distance limitations with clustering solutions. You will need to see if this fits within the capability requirements for the system concerned.

I have also previously mentioned server and storage consolidation. Obviously neither of these initiatives are particularly good in terms of geographic diversity but they do have other benefits in continuity assurance terms. Clearly you do not consolidate a production and resilient system on the same server but you can consolidate resilient systems together. You may leave your production systems as open systems, that is, individual servers per system, and consolidate the resilient nodes on a consolidation platform. A risk assessment has to be carried out here. Leaving production alone and not consolidating here can be a good idea as the activity would in itself be high risk but the risk is considerably lower when we consider building a resilience capability from scratch.

The risk is lower because we are not impacting the production environment to the same degree but there are still risks. Having a different architectural base to your high-availability environment from your production environment introduces an inherent risk. Changes applied to the production environment may need to be applied differently in the high-availability environment, or the impact on the overall environment of the change may differ. This introduces an added level of complexity in change management and design assurance. As a general rule, you should aim to replicate your production environment exactly in any secondary environment such as high-availability or disaster recovery. If you are unhappy with your architecture and think you can optimize it then take the time to do this in production and replicate the architectural change in your secondary environments. This is the idea of the activities at level 5 of this methodology, to prepare your production environment. If you do not do this you may find that in creating your new environments you have replicated all the problems that you have in your production environment. Where you may have coped with these problems in production you now have to manage them across two or three environments. You may find this unmanageable.

Moving from one site to a multisite environment architecture is not a trivial exercise. You may need to change your base architecture to take account of this fact.

9.6   People and Process

Resilience capabilities are not just about IT systems. We have already identified the critical people within your organization. We have discussed the need for a backup resource for each of these critical roles. This backup resource constitutes a resilience capability for the role.

Just as a system may fail or become unstable so to people are prone to illness and any number of other factors that may draw them away from work. In these events you need to ensure continuity. This may seem obvious but in my experience this is poorly done. In many instances a particular key person being away from work causes a number of business processes to stall because of no one knowing exactly what the status of tasks is, or no one is willing to make decisions in the absence of that person.

This is a difficult problem to address and the various tactics employed within this methodology hopes to address this problem. Having an organizational structure that lends itself to flexibility and proactivity helps to form a foundation for better continuity of resource. Having shadow resources will also help, as will managers with the experience to perform any role within their team. Adequate work management systems allow staff to see the progress of tasks and give them historic reference on decisions made and a frame of reference, and precedence, on which to base future decisions.

Processes should be written in a role-based manner. People should be referred to by their role, for example, the configuration administrator, rather than by name. An escalation matrix should be in existence, which can be utilized for both escalation and for the identification of secondary resource.

For business processes to continue to function we need both the people and systems that enable the process to be available and operational.

It is important when resourcing to choose people that are able to multitask. "One person one role" is a thing of the past and is inherently inefficient. The best scenario in continuity terms is to have a small number of highly skilled workers rather than a large number of button pushers.

Where possible and practicable, automate processes but try to leave yourself a manual workaround. Try to remove any reliance on low-level human interaction and leave your people to perform the skilled roles that they are required for, and to solve problems when they occur.

A story comes to mind of an associate who was working on an engineering process. It goes something like this. They were modernizing a plant that made say, widgets. They were unable to reduce staff levels too much so they took to retaining some of the manual processes and interlacing them with automation. The widgets came off the production line and into a bucket. There was a person that looked to see when the bucket was full and then pressed a button to empty it into a unit that processed them further. The problem was that everything went swimmingly except that the person on the button kept forgetting to press it and the bucket would overflow and cause the whole system to stop. In the end they programmed the system such that the bucket emptied itself. They left the button and the operator in place. The operator still kept pressing the button, when he remembered to.

9.7 Associated Benefits

There are some other benefits of having resilience for systems that I have not yet mentioned.

Where capability is cold standby the equipment only needs to be available, it does not need to be loaded or even booted. This equipment may be used for development purposes, or testing. Bear in mind that the disks will be immediately wiped if the system is needed by production.

When changes are being implemented it is prudent to first implement the change on the resilient system and ensure that there are no problems. If the system is load balanced switch the load to one node and implement on the other one first. If the change is implemented and tested okay, then switch the traffic to the second node, wait to make sure there are no additional problems, then implement on the other nodes. In this way we can make changes without any production downtime, that is, no interruption to service and the continuity of your business. The same applies when conducting hardware upgrades.

Figure 9.2 shows the primary node as the data master, replicating data to the resilient node. The resilient node is the system master. Changes are made to the resilient node before being applied to the production node. It is possible to automate replication of these system changes but it will generally be done via the configuration management server, as previously discussed.

Where distributed processing is employed it may be possible to move some physical services locally to the system and provide some added benefit for the

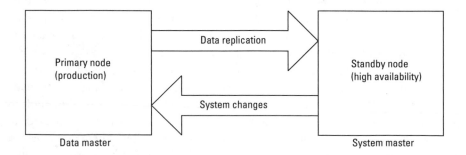

Figure 9.2 Systems configuration management.

users or customers. There may be some benefit to be drawn from locating the systems local to a particular user group or customer base. These could simply be access-speed related.

It should be said that high-availability capabilities are always on a system-by-system basis. That is, the capability will consider an isolated failure of that system alone. The resilient system, where required, will interface back to the production systems that are still operational. The failover of one system does not necessarily imply the need to failover other systems. This in not usually the case where we consider general disaster recovery scenarios.

9.8 Security and Safety

Just as we may implement resilience capabilities around people and process, so too can we for security and safety.

Ensure that you have adequate numbers of security personnel to fill any gaps that may occur due to sickness or holiday. Ensure that your personnel levels are such that guards work short shifts. Typically, security guards work very long shifts, which is illogical. Some guards work 12 hours straight. It is impossible to maintain the level of attention required for this job over this period. Shifts should be half this figure. Ensure that staff are rotated on role and location to provide variation and thus interest. This will also ensure that staff have an ability to perform a number of roles and are familiar with the layout of all locations. This is imperative to providing a coordinated response in a crisis situation. Each security staff member needs to be able to visualize the position of the other members of the team when a situation occurs.

In terms of safety we have already put in place the safety equipment required. At this level of the model we should ensure that we have some spare backup equipment to be used if some of the equipment fails or is damaged. This should be stored at locations across the organization and wherever there is a concentration of equipment. For example, if there are 50 fire extinguishers in a building then there should be another two or three spare ones located somewhere within the building. If, on inspection, an extinguisher is found to be faulty it can be replaced immediately.

Where outsourcing is used for some security and safety services you should consider having a backup agreement with another supplier to be used in the event that the primary supplier suffers an interruption to their service.

9.9 Level 6 Performance Indicators

Table 9.3 sets out the performance indicators at this level of the continuity assurance model.

Table 9.3
Level 6 Performance Indicators

Resilience	
Item Number	**Action/Task**
R6-001	Review applications architecture in terms of suitability to replicate and impact on data consistency. Take action to remediate if possible and practical.
R6-002	Assess replication technology in terms of potential data integrity issues within your environment.
R6-003	Initiate a data architecture review with a view to rationalize where appropriate. Determine critical data to be stored and replicated.
R6-004	Confirm the existence and operation of a time-server within your infrastructure. Remediate where necessary.
R6-005	Design and implement a fail-over sequence and associated technology for distributed systems.
R6-006	Assess the costs, benefits, and feasibility of implementing server or storage consolidation in the high-availability environment only, without additional change to the production infrastructure (optional, rate the assessment if no implementation occurs).
R6-007	Determine a resilience strategy in terms of location of your high-availability infrastructure.
R6-008	Implement resilience capabilities for critical systems based on the criticality classification of the systems.
R6-009	Ensure that critical resources have a degree of resilience capability by the allocation of secondary resources that could perform their roles in a crisis situation or situation where the primary resource is unavailable. Ensure that a general escalation matrix is in place for all critical processes.
R6-010	Review the state of any work management system that exists within the organization. Ensure that tasks associated with any critical process are documented and tracked.
R6-011	Assess workforce requirements across the organization with a view to rightsizing and automating low-level processes. Address any training requirements born out of this exercise (optional).
R6-012	Review all processes and harmonize for a multienvironment landscape. Review overall environments architecture including development and test environments with a view to optimizing the use of hardware across all environments.

Table 9.3 (continued)

Resilience	
Management Functions	
Item Number	**Action/Task**
M-023	Ensure that all changes, including the commissioning of new equipment, are managed through the change control process and are reflected in the configuration repository.
M-024	Ensure activities at this level are adequately communicated across the organization.
M-025	Continue governance of projects commissioned at earlier levels.
Security and Safety	
Item Number	**Action/Task**
S1-017	Review shift structure and resource levels for physical security staff.
S1-018	Assess the feasibility of backup outsourcing agreements for security-related services, should additional resource be required in a crisis situation. Implement where required (optional).
S1-019	Considering the high level of change in technical environments at this level, general system and network security should be regression tested when other work at this level is complete. Issues should be remediated.
S2-012	Procure and implement backup security equipment as required.
S2-013	Assess the feasibility of backup outsourcing agreements for safety-related services. Implement where required (optional).

10

Level 7: Recovery

Recovery is your last resort in terms of maintaining an acceptable level of continuity. Should all else that you have put in place fail this is what you will be left with. It is a last resort but for some types of events it will be your first action.

Here we are concerned primarily with recovery from a disaster event. A disaster can be defined as an unplanned event that severely impacts the continuity of the organization and whose effects are widespread, affecting multiple people, processes, and/or systems.

Due to the widespread impact of a disaster it is necessary to have a disaster recovery site, or sites, located away from standard company facilities. This is the case as many disasters are centered on a particular location.

The purpose of a disaster recovery facility is to provide the capability to recover critical components of your organization. These critical components have been determined at previous levels of this model.

Components should be recovered within the target recovery time and point objectives, RTOs and RPOs. These objectives have been considered within the criticality rating process for systems and should be achievable utilizing the technology configurations set out for the particular levels of criticality.

It is assumed at this level of the model that you have already implement a high-availability (resilience) environment in line with the requirements of this methodology. You should have also concluded all work on process advancement required in previous levels of the model.

We are now faced with the task of integrating a third environment into your technology landscape, the disaster recovery environment.

10.1   Capability Configuration

Table 10.1 sets out the capability configuration requirements for resilience and recovery of critical systems.

The critical systems are categorized as primary, secondary, and ancillary, as previously defined.

We can see that the capabilities to be implemented for recovery are generally less advanced than the capabilities implemented for resilience. This is generally because the time and point objectives to be met for recovery are not as rigorous.

It is usually accepted that if we are in a disaster situation we have some tolerance to outage, be it business or customer tolerance. Additionally, as we will only use this capability in the most extreme circumstance, it is not practical to overspend on technology when we have a high-availability capability also in place that will take care of the majority of continuity events we encounter.

Of course the RTO and RPO will differ for each system and so to will the specific technology used to affect the capability. I can implement a warm-standby system with a 1-hour RTO or a 4-hour RTO, for example.

You will find the definitions for each of these capabilities in Chapter 9.

Note that while we do consider resilient servers as part of the production infrastructure, unless they are configured into a separate high-availability site, we clearly do not need any recovery capability for these servers. That is, if we have a system with one server and hence one resilient server, we only require one server for recovery, not two. We are not in the business of recovering a resilience

Table 10.1

Recommended Capabilities

Capability/Criticality	Primary	Secondary	Ancillary
Reciprocating load	Where no outage is acceptable.	N/A	N/A
Reciprocating data	Where data integrity is an issue.	Where data integrity is an issue.	N/A
Hot standby	Minimum capability for resilience.	Recommended.	N/A
Warm standby	Minimum capability for recovery.	Minimum Capability for resilience.	Recommended.
Cold standby	N/A	Minimum capability for recovery.	Minimum Capability for resilience, and recovery (only where system is dependent).

capability. This is the same when we consider a clustered, load-balanced, system. As long as one node can take the full load then we only need one node in the recovery site. This seems simple enough but you would be surprised at how many people do not get it.

10.2 Recovery Profiling

From the RTOs previously determined, we can build up a picture of a recovery profile. That is, an order of systems to be recovered in terms of shortest recovery-time objective to longest recovery-time objective.

This recovery profile needs to be looked at, and possibly fine-tuned, to take account of system dependencies. Remember that until now your assessment of recovery time and point objectives has been based on user or business tolerance. This now needs to be factored in terms of technical reality.

Technical dependencies between the critical systems may mean that the sequence needs to be altered. As such, the time objectives of any given system may need to be tempered with what is technically possible. In addition, you may identify additional systems that have not previously been identified as critical but are themselves critical by association, in that they are required to be part of the infrastructure because another critical system has a run-time dependency on them, or it.

For example, system A may have a RTO of 8 hours but system B has an RTO of 4 hours. This is fine but consider the case where system B interfaces with system A and needs this system available in order to function. Your process analysis may not have picked this up unless someone in the business knew of this dependency. You may have picked it up when doing the bottom-up criticality assessment of systems but it always pays to revisit these dependencies at this level, to make sure nothing is missed. In the example, system A will need to be moved in the recovery sequence so that system B can meet its RTO. Remember that an RTO is meant to be a maximum time objective; there is no issue with recovering a system ahead of time.

Where we have systems that have been assessed as noncritical, but find that they have associated critical systems that interface to them, we need to assess the nature of the interface. If this interface is part of a critical process then the system has been incorrectly assessed and needs to be looked at again in terms of criticality. If the interface is noncritical we need to look at the functionality of the associated critical system and determine the technical constraints around this interface. It may be the case that we need to make an application change to this system, or a change to the middleware infrastructure in the disaster recovery site to harness this interface. This means that we simply block this interface through a change in the application or interface code. If the interface mechanism is batch transfer, or similar, this may be simple. The application may not care if the interface is active or not. However, in some cases, particularly where the interface is real time, the interface may need to be harnessed. If you do not do this the application may "panic" or "hang" on the interface. You may find that the entire system comes to a halt.

10.3   Fail-Over Scenarios

When we consider invoking the disaster recovery environment we will typically consider a total-loss scenario only. For system-isolated failure we have our high-availability environment, which will be nearby or colocated with the primary production systems. When I say total loss only, I am referring to the loss of the primary facility as a whole or a significant portion of it. We may declare this facility to be lost under a number of circumstances. These circumstances could be:

- Loss of a service that affects all, or a significant portion, of the systems in the primary facility;
- An event that affects a considerable portion of the facility and/or has the possibility of spreading throughout the facility;
- A disaster event that affects the facility as a whole;
- An event is anticipated that warrants preemptive recovery;
- An event has occurred nearby and operations are moved to the recovery center as a precaution.

Clearly, in most circumstances the high-availability site will be used, unless this site is colocated within the primary production facility. There are many types of events that will, or could, have an impact on the general area of the facility. In this case, remembering that the high-availability facility will be nearby, you would utilize the recovery site. You may transfer operation directly to the recovery center or you may operate for a short time at the high-availability site and then move to the recovery center.

Remember that the high-availability site can be recovered to faster, and on a system-by system-basis as required. The high-availability site should support incremental failover of individual systems.

When I refer to incremental fail over, I am talking about the fail over of one or many systems to their resilient nodes. It is not necessary to fail over all critical systems to their resilient nodes at the same time. This is because the high-availability environment is configured such that the systems can interface back to the primary production node of any connected systems. This is possible because of the close proximity of the high-availability site and the network technology used. Your disaster recovery site is unlikely to have equivalent capabilities.

You should also consider any interfaces back to production systems that are not critical. In a high-availability situation the resilient system will pick up these interfaces back to other systems that exist in the production environment but do not exist as part of the high-availability infrastructure. These interfaces may not be critical, indeed if they were the systems would probably be critical also, as previously discussed, but the interface will not be harnessed on the high-availability system. The system will function exactly as if it is the production node and will expect the interface to be there. This is because high-availability is focused on individual failure of particular systems. It is not designed to cope with the total loss of the primary production facility. This is why we have a separate disaster recovery environment and indeed why in many cases resilient systems are colocated with the primary production system. Remember that the high-availability environment is merely an extension of the production infrastructure, whereas, the disaster recovery environment is self-contained.

When a disaster is declared we move operation of all critical systems to the recovery platform simultaneously, or at least close to it while following the recovery sequence. It is not usual to move operation of a single system to the recovery site. Depending on the technology involved this may be possible and may be actioned where a failure has occurred to both the primary production system and its resilient node. This should be done on a needs basis. Procedures for actioning this could be detailed as part of the contingency plan for each system.

A disaster recovery plan should be defined at this level of the continuity assurance model and should include detailed procedures for fail-over of all critical systems according to the recovery sequence.

Whereas fail over to a high-availability system will mostly be automatic the fail-over to the recovery environment will often be a manual process. This is clearly driven by the particular capabilities implemented.

There are a number of ways to affect this fail-over. Some are listed below. Which method you employ will depend mostly on your particular network topology:

- Redirection of traffic through a change in the primary router configurations;
- Change of IP addressing of recovery systems to match production IP addresses;
- Configuration of an alternate routing configuration into the network for a largely automated solution.

We have spoken at length about replication mechanisms in Chapter 9 on resilience so I will not cover this again in detail here. However, you will need to decide on the method of replication of critical data. This will largely depend on what infrastructure you have put in place at previous levels of the model.

If you have implemented a SAN infrastructure then replication for disaster recovery will be much simpler than otherwise. Note that due to the reasonable distance between the production and recovery sites it is usual for replication to be asynchronous for recovery purposes. This depends of course on the RPOs for the critical systems concerned. If you intend to utilize media such as tape for recovery be sure to calculate exactly how long this will take and ensure that you can meet the RTOs and RPOs. Ensure that the tape units at the recovery site match those in the production site exactly as most media requires specialist or proprietary equipment. Be sure that you know exactly how long it is likely to take to load the data. It can take a lot longer than you may think to restore from tape.

When recovering Wintel based systems be very careful with the use of recovery tapes. If you are recovering the operating system as well as the applications be aware that you will be restoring all the Windows drivers from the production system. This will include screen, disk, and general I/O drivers. If your hardware in the disaster recovery facility does not exactly match that in production you could have a problem. This is because you may be loading incompatible drivers for your hardware. Even if your hardware has the same model number as the system in production the internal components used in that particular batch could be different. For this reason it is generally not a good idea to restore Wintel operating systems from tape. It is possible to encounter similar problems with some combinations of Unix operating systems and backup/restore applications.

Looking at replication on a system-by-system basis can be problematic due to the large range of technology options. Each system vendor will have their own preference for replication and you could end up with a number of different technologies operating across your infrastructure. This is not optimal in terms of design assurance but is sometimes the only alternative.

Remember that the degree of complexity of your architecture and recovery process is inversely proportional to the probability of successful recovery. Where possible, keep it simple.

10.4   Disaster Declaration

There are four categories of disaster declaration. These are:

1. Active;

2. Preemptive;

3. Precautionary;

4. Planned.

The crisis management team is responsible for disaster declaration. Once a disaster is declared, responsibility falls to the continuity assurance group to action the recovery procedures. The crisis management team remains in governance of this activity.

Active recovery is recovery declared in the midst or aftermath of a disaster event. Disaster events should be categorized and detailed at this level of the model.

Preemptive recovery may be declared in preparation for an event that is going to occur or has a very high probability of occurring imminently.

Precautionary recovery is declared when a situation occurs that has the possibility of developing into a disaster or there is intelligence received that would indicate a possible event. This is different to preemptive recovery as there is not necessarily any timescale for the occurrence of the event. Organizations may move to the recovery site for a longer period of time until the perceived threat has passed. An example may be a general terrorist threat on the company or a localized threat to the area in which the production facility is situated.

Planned recovery may be initiated for a range of other nonthreat-related reasons. This may include testing or could be actioned while renovation work is undertaken at the primary production facility, for example.

When a declaration occurs there should be detailed procedures in place to be followed. These recovery procedures, including the declaration procedure itself, will be governed by and referenced within the organizations crisis management plan.

10.5 Fail Back

In focusing on the creation of a recovery capability and the fail over to a disaster recovery facility, we can often forget about the fact that at some point we have to fail back to the production environment. This could be the same production environment or it could be a new environment that has been built because the original one was destroyed in the disaster event.

In some ways fail back can be more complicated than fail over. This is because you could be failing back to an environment that is not clean, that it, it contains residual data from before the fail over. You could possibly purge the systems of data and begin from scratch but this may involve restoring first from tape and then merging in the data from the recovery platform. This is because your production environment may contain a large amount of on-line data that was not loaded in the recovery environment because it was not required for critical processes, but is required in the business-as-usual production environment. Either way, you will need to merge the operational data generated while the disaster site was running as production back into the full production environment.

Again, this is a discussion about replication, data integrity, and so forth, and we will not go into this in detail again as it has been covered previously. Be aware though that you will probably have to bring the entire environment to a stop before switching back to true production. You should be able to utilize the recovery sequence again and you should also only need to synchronize new or modified data, reducing bandwidth requirements.

10.6 Environment Management

It needs to be recognized that you are now working a multienvironment, multisite technical landscape. This complex environment structure requires careful management.

I would like to take a moment to define the different types of technical environment that exist within a typical organization. These are, in no particular order:

- Production;
- High availability (resilience);
- Disaster recovery;
- Software quality asurance (SQA);
- Integration test;
- Integrated development environment (IDE).

Some organizations have more test or development environments, but I have found this to be the most efficient environment model. Note that the SQA environment is sometimes called system test, preproduction, or staging, depending on the nature of software development and integration conducted within your organization.

In an ideal world these environments will be spread across four physical locations: production, high availability, disaster recovery, and development and testing. It is not a good idea to have development and testing colocated with production or high availability; however, colocating with the disaster recovery environment could be an option, but the physical location must be set according to disaster recovery requirements, and thus may make this impractical.

You will need to determine a promotion model for application of system changes, deployment of new software, and other events. You will also need to consider the replication of data across the potential production environments. Following on from our discussions in Chapter 9, we arrive at the following, represented by Figure 10.1.

Note that you may not be allowed production data in your SQA environment for information security reasons. Figure 10.1 however shows replication to this environment. The integration test and IDE environments utilize test data. Static data should be treated in the same way as a software code change. This promotion model should be managed by the configuration management function within your organization.

We previously had the high-availability environment as the system master and the production environment as the data master. We now change this to incorporate the disaster recovery environment, which now becomes the new system master. System changes are applied first to the recovery environment, and then to the resilience and production environments, in turn. Be aware that if the change works in preproduction environments and then fails when implemented in production the change needs to be backed out in all environments and redesigned. This situation may also highlight possible consistency problems between your environments.

Depending on your application architecture it may be possible to merge the SQA, integration test, and IDE environments onto a single infrastructure. This is very cost effective in terms of keeping hardware costs to a minimum. It may even be possible to utilize part of the recovery infrastructure for this purpose, that is, the part(s) that are implemented in a cold standby capability configuration.

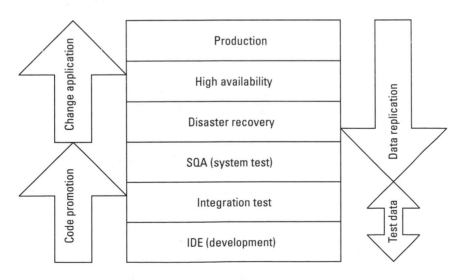

Figure 10.1 Technical environments management.

We achieve this by creating multiple discrete execution environments across the same hardware infrastructure. An execution environment is a set of discrete processes running on each server that constitute the complete set of applications required in the environment. They communicate with each other over a unique allocation of interfaces. For example, if we have four execution environments on a system then we will have four separate instances of the same application running on that system in parallel. By connecting all the interfaces for each application across a different set of communication ports we achieve four complete execution environments across the same infrastructure. This is complicated to achieve and you will obviously need to up the capacity on your systems to cope with this, but it is a very efficient way to manage your nonproduction environments. Each application instance runs from a different area of disk where an entire set of binaries and/or code resides. In this way each environment can be running a different version of the application, with different dynamic data if required. This of course may not be possible; it all depends on the architecture of the applications concerned. This type of combined environment is what I call a multithreaded execution architecture (MEA).

Note that the existence of a SAN should not impact your ability to achieve this, in fact it may help. If you are running on a consolidation platform this sort of clever configuring is not required. The consolidation technology will give you the same capability, without the headache.

10.7 License Management

There may be significant implications to your existing software license agreements in establishing new technical environments.

You will need to conduct a review exercise across all existing software agreements, operating systems, applications, management tools, everything. In many cases you will find that the license agreement you have will not mention anything about use in recovery. If this is the case you will need to approach each of the software vendors and notify them of your intent to utilize a secondary, license in one or more recovery environments (high availability and disaster recovery). I find that the best way to negotiate this is to tell them what you are going to do and ask them to let you know, within a certain time frame, if they have a problem with this.

Traditionally, most companies allowed some degree of use for recovery but this has changed in recent years. You may find that you have some significant costs in obtaining recovery license agreements for your software. If you have an enterprise level agreement you will probably be okay, and may only need to pay a small fee, if at all.

Unless your license agreement specifies the exact hardware that the system is to be run on you will be entitled to run one instance of the software, for each license you have, on any hardware. This means that you can run the system in the recovery or high-availability environment as long as the production system is shut down. This is great but does not account for situations where you may have to run the systems in parallel, the most common situation being capability testing. This said, there are many vendors who will allow a certain number of days per year that you may run dual licenses for testing purposes only. You may need to get a letter from them to substantiate this though as it is not always detailed in the license agreement.

Be aware that some software requires a license key supplied by the vendor in order for you to load or run the software. These license keys, for more expensive software, are often keyed to the host-id of the system that you are running the application on. They are not transportable. If you want to load the software on a different server you will need to approach the software vendor with the details of the system to ask them for a new license key, and give an explanation of why you should not be buying another license.

10.8 Location

Selecting the physical location of your recovery facility is an important task and one that requires a good deal of thought. You will need to look at your residual risk profile and determine the magnitude, in terms of distance from your primary facility, of any potential event.

It is generally accepted that a recovery facility should be at least 60 km from your primary facility. I do not really subscribe to this view. As long as you are far enough away to avert any potential event that you have profiled then I think the closer the better. Remember that you will have to move people in the event of disaster. The recovery site will need to be staffed with technical personnel of which at least some will have to travel from the production facility. You need to take into account transport links and, if utilizing public transport, the frequency of departure. Unless your operation is based in a reasonably large city you may find that to get more than 60 km away and have the supporting facilities you require you may need to move to the next city. This may mean a plane trip for any personnel that have to relocate. This is not generally a good idea as airports can be difficult places to navigate given that you may have just experienced a regional disaster.

Make sure that your location has local facilities such as car parking, restaurants, and hotels to cater for your staff. Exactly what facilities you can cater for yourself will largely be determined by whether you select to outsource your recovery facility or acquire a facility yourself.

If possible, move operations to your high-availability site until your staff have had a chance to relocate to the recovery facility, if indeed your high-availability site is still available. Even if some processing is possible at this site it may act to minimize the impact to your customers while you initiate a wider recovery operation.

10.9   Workplace Recovery

When we speak of recovery we are not only concerned with the recovery of IT systems. Depending on the nature of the disaster that has occurred you may need to relocate staff to alternate office space to ensure continuity of business process.

When we talk about workplace recovery it should be noted that to be able to operate as normal, staff will need access to basic facilities such as computers, printers, telecommunications, and so on. They will also need to be located somewhere that can be easily accessed from their primary workplace, and that has amenities such as restaurants and the like available within the complex, or nearby.

Note that a workplace recovery facility does not need to be colocated or even in close proximity to your technical recovery environment.

There are a number of options that every organization has for workplace recovery. The main options are listed as follows:

- Utilize another operational office space within the organization that is involved with noncritical processes.

- Put in place a contract with one of the many workplace recovery facility providers.

- Put in place a contract with a provider of serviced offices.

- Utilize a facility that is likely to be available and can be altered to become fit for purpose. The most common example here is to use an indoor sporting facility, something with a lot of floor space where workstations can be erected. You will need to have equipment permanently stored on site that can be quickly setup. Many companies that have their own sporting facilities and make use of them in this way.

Again, you will need to decide whether you are going to outsource or insource this capability. We will look at the costs and benefits of insourcing and outsourcing in a later chapter.

It is important, and a requirement of this methodology, that workplace recovery is considered as a whole across the organization. It happens too often that each department within an organization is left to make their own plans in this matter.

10.10 Testing

Testing is imperative to the ultimate success of any disaster recovery capability. Testing of the recovery facility and recovery plan should be conducted as regularly as is practical, but in any event, a full test should be conducted once every year. This test should include the key people that would be the main players in any real recovery operation.

Because of the potential impact to the production environment should the test go wrong, we limit most tests to a simulated fail over. The systems are synchronized and brought on-line within an isolated segment of the network. A simulated fail-back may also be possible and is recommended if the impact of such a test is acceptable.

It is important to conduct the test as if it is for real. All steps in the process should be addressed, even if the step is simulated to some degree. External security agencies and emergency services should be involved to the degree that is required to prove the process.

When creating the processes and technical capabilities a rigorous series of tests is required. If we assume that you are building the recovery infrastructure from scratch, the following layers of testing need to be worked through. These are:

1. Equipment audit;

2. System test;

3. Integration test;

4. End-to-end test;

5 Tabletop process test;

6. Network security test;

7. Disaster recovery fail over test.

As equipment arrives on site it is important to ensure that it meets the specifications requested of the supplier. This stage is the *equipment audit.*

System test is a technical and functional test of each individual system in turn to ensure that it behaves the same as the production copy. Remember that in a disaster recovery implementation we are not implementing new systems, so testing is not quite as rigorous as if this was a new system being integrated. All that we need to prove is that the system functions in the same way as it does in the primary production environment.

Integration test is simply looking at each system in turn and testing the functionality of the interfaces. Test scenarios should be run that utilize these interfaces.

End-to-end test, as the name implies, tests the complete functionality across the entire environment. This is a functional test and is also intended to prove that the environment as a whole behaves in the same way as the primary production environment. It is useful to define scenarios that enact the critical processes through the environment. If possible, utilize production data for the test and compare the outputs from the disaster recovery environment to that of the production environment.

At this stage we begin testing the process side of the capability by conducting a *tabletop process test*. This test is performed by gathering representatives from the key areas of the organization that will be involved in any recovery exercise and walking through the recovery process. The participants are given an event scenario and have to walk through on paper the tasks that they perform at each step of the recovery sequence. This test includes any decision-making process that may be required as a consequence of the simulated event.

In terms of security testing, we are essentially replicating the production environment like for like, so the same security that exists in production should exist here. For this reason it should not be necessary to conduct any specific security tests on the disaster recovery systems, other than to substantiate that the configuration is the same as that in production. As the network in the recovery environment will be newly designed and implemented, a degree of *network security testing* should be performed.

Finally, we combine the physical fail-over of systems with the recovery process to conduct a simulated *disaster recovery fail over test*. Once the fail over has been actioned, a regression test of end-to-end scenarios is performed and results compared to the output from the production environment. As part of this test the technical and procedural aspects of the fail back operation should also be tested. The test plan and scenarios used at this level of testing should be used as the basis for ongoing testing of your disaster recovery capability.

Be sure that test plans and scenarios are updated in line with any changes to the environment that have an impact on them.

A degree of performance testing should be performed as part of the disaster recovery fail over test. It is usually acceptable for a disaster recovery site to be slightly behind the production environment in terms of performance, but this will depend entirely on the requirements of your business. It is important to consider this and determine some benchmarks with which to assess the performance of your environments.

In some cases, where the technology being used, particularly for replication, is new or in some doubt, it may be useful to conduct a proof of concept exercise as a test of this technology before embarking on implementing a full-scale recovery capability. The layers of testing described above are enacted similarly for a proof of concept test as they are for a full scale capability test, but with some minor differences. If we are prototyping only two systems the integration test and end-to-end test can probably be combined, as they will be the same, or nearly the same, test. There is no need to test the process side of the capability assuming that the proof of concept infrastructure is not intended to be used as a real disaster recovery platform in itself.

10.11   Security and Safety

Clearly the recovery facility must meet the same standards as the production center in terms of security and safety. In terms of the IT network infrastructure, testing should be carried out to ensure that it is secure, as previously discussed.

In a disaster situation particular attention should be paid to the well-being of staff and their families. It should be possible for people to move to a recovery site in safety, so efforts should center on the safety of transport.

If you are suffering the effects of a generalized disaster that has affected your entire city or region then special care must be taken. You should have plans in place to safeguard not just your key staff members but also their families. In such a situation the first thing many people will do is rush back to their families to make sure that they are all right, which is quite understandable. Be prepared to relocate not just staff to the proximity of the recovery site but also their families. Ensure that transport is arranged for them to lessen any worry. Remember that these people are the fabric of your organization, without them you can have all the technology in the world and your critical business processes will still falter. You can bind people to do all sorts of things contractually but contracts quickly go out the window when the safety of one's family is in question.

People in general want to work and want to do the right thing by their company; you simply need to put them in a position where they can do this safely. We are not only concerned for key staff at time of recovery but for all staff in returning to work after the event has passed. The assumption of course is that noncritical staff are allowed to go home in a disaster situation. This is one course of action and may increase the general safety of the workforce depending on the nature of the event. However, you do not want these people to linger away from work after the situation has passed because of fears they have over their personal safety. Postdisaster, you need to ramp-up security and take closer note of the well-being of the your staff. Ensure that they are all properly debriefed so that they understand the event that has occurred and understand any future threat. Understanding is the key to recovery. Be aware that some may be in a state of trauma or shock. Ensure that they receive care and attention. A person whose mind is not on the job is a liability in continuity terms. You do not want a secondary event occurring as a result of poor return from the initial event.

10.12   Level 7 Performance Indicators

Table 10.2 sets out the performance indicators at this level of the continuity assurance model.

Table 10.2
Level 7 Performance Indicators

Recovery	
Item Number	**Action/Task**
R7-001	Construct a recovery profile (timeline) based on RTOs.
R7-002	Determine performance objectives for the critical systems in the disaster recovery environment.
R7-003	Review critical systems and construct a dependency diagram both for run-time and startup dependent interfaces. Modify your recovery profile as required.
R7-004	Review interfaces to noncritical systems and determine a strategy for harnessing these interfaces on the critical systems in the disaster recovery environment.
R7-005	Determine the possible scenarios, procedures, and technical mechanisms for replication and fail-over to the recovery environment. Develop the disaster recovery plan(s) on the basis of agreed strategy.
R7-006	Revise system contingency plans as required.
R7-007	Determine the criteria for classification of a disaster event. Determine the process for disaster declaration.
R7-008	Update crisis management team terms of reference, crisis management plan, and emergency action plan as required.
R7-009	Define and document an organization wide strategy for the management of technical environments.
R7-010	Determine a data replication sequence and process for replication of data between environments.
R7-011	Determine a promotional model for use in propagating software between environments. Update change and configuration management procedures as required.
R7-012	Review software license agreements for all critical systems to determine any restrictions for recovery use. Assess requirements and negotiate with software vendors. Procure additional licenses where required.
R7-013	Determine technical recovery center location strategy in terms of target distance from primary facility and whether the provision of the facility should be outsourced or not. Assess possible recovery site locations and facilities for risk and benefit and concluded the decision making process.
R7-014	Determine workplace recovery requirements in terms of numbers, facilities, and locations. Decide on insourcing or outsourcing. Conclude decisions, and contracts where applicable.
R7-015	Determine implementation strategy and plans. Commission projects where required.
R7-016	Determine testing strategy, plans, scenarios, and scripts. Implement supporting tools where required. This task includes plans for ongoing testing.
R7-017	Implement disaster recovery capabilities for critical systems in line with criticality classification and requirements under this methodology.

Table 10.2 (continued)

Recovery	
Item Number	**Action/Task**
R7-018	Execute testing to prove capability.
R7-019	Perform final updates to all related or affected processes and procedures.

Management Functions	
Item Number	**Action/Task**
M-026	Ensure that all changes, including the commissioning of new equipment, are managed through the change control processes and are reflected in the configuration repository.
M-027	Ensure that all activities, especially new processes, are communicated across the organization.
M-028	Ensure that all projects are concluded and properly closed out. Initiate a minimum of 5 spot-quality reviews at this level in addition to a full audit. This audit should be conducted by an external agency, security permitting.
M-029	A management process should be established for the dissemination of information and debriefing of personnel in the wake of a disaster event.

Security and Safety	
Item Number	**Action/Task**
S1-020	Audit technical and workplace recovery facilities to ensure compliance to security standards. Remediate where required.
S1-021	Review security relating to transport between primary and recovery facilities.
S1-022	Implement plans for increases in security both throughout and after the disaster event, to abate fears of staff returning to work.
S2-014	Audit technical and workplace recovery facilities to ensure compliance to safety standards. Remediate where required.
S2-015	Review safety relating to transport between primary and recovery facilities.
S2-016	Determine a strategy for safety of key personnel's immediate family in the event of disaster. Implement plans to affect this.
S2-017	Implement plans for the movement and location of all staff in disaster events. Determine who will be sent home and who will stay. This plan will affect critical and non-critical personnel alike.
S2-018	Post event, ensure that care is provided to those suffering from trauma or shock related to their experiences of the event. Put a plan in place to ensure that this is done.

11

Continuity Assurance Achievement Rating (CAAR)

This chapter is concerned with the overall rating of continuity assurance capability. At each level of the model I have listed a number of performance indicators. These are represented by tasks that need to be actioned in order to attain recognition of some accomplishment at that level of the model.

Here we are concerned with rating or scoring your performance on each of these tasks. It should be noted that we do not only rate how well you performed the actual activity, we are concerned primarily with the results in terms of increasing your organization's continuity assurance capabilities by adherence to the model. It follows that if you have performed the task poorly, then the resulting gain to your organization will also be poor.

11.1 Rating Matrix

The tables set out below represent the key performance tasks at each level of the continuity assurance model. The full description of the task has not been included here; instead, I have summarized the nature of the activity in a few words. You should refer back to the individual levels of the model for the complete descriptions.

You will note that there are a small number of tasks that are optional. These represent certain aspects of the model that are recommended but not deemed absolutely essential, in that the core principles of continuity assurance can be achieved without performing these tasks. In some cases optional tasks concern the implementation of certain types of capability technology that may not be applicable, or suited, to your organization. Note that optional tasks have been accounted for in the overall scoring regime in terms of the setting of capability thresholds at each level of the model.

Any task listed as nonoptional is quite clearly mandatory. The task must be attempted or addressed in some way.

A success rating is given to each task, and is intended to represent your achievement in performing the task and in gaining increased capability as a result. Rating is largely qualitative in analysis, but is represented as a quantifiable number from zero to five. The rating levels can be defined as set out in Table 11.1.

Table 11.1
Success Rating Matrix

Rating	Task Status	Success
0	Incomplete	Task not started, performed badly and/or no benefits.
1	Incomplete	Largely unsuccessful with minimal benefit derived.
2	Incomplete	Poor results in general but can be built upon.
3	Complete	Task has been performed adequately.
4	Complete	The task is successful with room for improvement.
5	Complete	Task is complete and highly successful.

Level 1

Capability scoring at this level may be recorded utilizing Table 11.2.

Table 11.2
Level 1 Capability Scorecard

Item	Level	Summary	Comp.	Opt.	Success Rating	Weight Cat.	Max. Score
R1-001	1	Continuity assurance team	Y/N	N	0–5	3	15
R1-002	1	Rationalize organization	Y/N	N	0–5	1	5
R1-003	1	Communication pathways	Y/N	N	0–5	2	10
R1-004	1	The blended team	Y/N	Y	0–5	1	5
R1-005	1	Roles and responsibilities	Y/N	N	0–5	1	5
R1-006	1	Management processes	Y/N	N	0–5	2	10
R1-007	1	Skills review	Y/N	N	0–5	1	5
R1-008	1	Management tools	Y/N	N	0–5	1	5
R1-009	1	Information gathering	Y/N	N	0–5	1	5
R1-010	1	Commission projects	Y/N	Y	0–5	1	5
M-001	1	Communication road map	Y/N	N	0–5	3	15
M-002	1	Governance model	Y/N	N	0–5	3	15
M-003	1	Problem management	Y/N	N	0–5	1	5
M-004	1	Continuity assurance strategy	Y/N	N	0–5	3	15
M-005	1	Set and allocate budgets	Y/N	N	0–5	1	5
M-006	1	Security/safety gap analysis	Y/N	N	0–5	1	5
M-007	1	HR process review	Y/N	N	0–5	1	5
S1-001	1	Security integration	Y/N	N	0–5	2	10
S1-002	1	Remove subcontractors	Y/N	Y	0–5	1	5
S1-003	1	Security review	Y/N	N	0–5	1	5
S1-004	1	Security roles and responsibilities	Y/N	N	0–5	1	5
S2-001	1	Health and safety integration	Y/N	N	0–5	2	10
S2-002	1	Safety roles and responsibilities	Y/N	N	0–5	1	5
S2-003	1	Safety information	Y/N	N	0–5	1	5
Total:							180

Level 2

Capability scoring at this level may be recorded utilizing Table 11.3.

Table 11.3
Level 2 Capability Scorecard

Item	Level	Summary	Comp.	Opt.	Success Rating	Weight Cat.	Max. Score
R2-001	2	Vulnerability analysis	Y/N	N	0–5	2	10
R2-002	2	Vulnerability standard and process	Y/N	N	0–5	2	10
R2-003	2	Remediate vulnerabilities	Y/N	N	0–5	3	15
R2-004	2	Risk assessment	Y/N	N	0–5	2	10
R2-005	2	Risk mitigation	Y/N	N	0–5	3	15
R2-006	2	Residual risks register	Y/N	N	0–5	2	10
M-008	2	Communication and training	Y/N	N	0–5	2	10
M-009	2	Documentation	Y/N	N	0–5	1	5
M-010	2	Cultural change	Y/N	N	0–5	1	5
M-011	2	Project governance	Y/N	N	0–5	2	10
S1-005	2	Security vulnerability analysis	Y/N	N	0–5	2	10
S1-006	2	Security standards	Y/N	N	0–5	2	10
S1-007	2	Identify and mitigate security risks	Y/N	N	0–5	3	15
S2-004	2	Safety vulnerability analysis	Y/N	N	0–5	2	10
S2-005	2	Safety standards	Y/N	N	0–5	2	10
S2-006	2	Identify and mitigate safety risks	Y/N	N	0–5	3	15
Total:							170

Level 3

Capability scoring at this level may be recorded utilizing Table 11.4.

Table 11.4
Level 3 Capability Scorecard

Item	Level	Summary	Comp	Opt.	Success Rating	Weight Cat.	Max. Score
R3-001	3	Business impact assessment (BIA)	Y/N	N	0–5	3	15
R3-002	3	Systems criticality rating (SCR)	Y/N	N	0–5	3	15
R3-003	3	Capability charts	Y/N	N	0–5	3	15
R3-004	3	Risk/criticality relationships	Y/N	N	0–5	3	15
R3-005	3	Risk chart/critical risks	Y/N	N	0–5	3	15
M-012	3	Socialize results	Y/N	N	0–5	3	15
M-013	3	Quality review	Y/N	N	0–5	2	10
M-014	3	Documentation	Y/N	N	0–5	2	10
M-015	3	Project governance	Y/N	N	0–5	2	10
S1-008	3	Assess critical security risks	Y/N	N	0–5	3	15
S2-006	3	Assess critical safety Risks	Y/N	N	0–5	3	15
Total:							150

Level 4

Capability scoring at this level may be recorded utilizing Table 11.5.

Table 11.5
Level 4 Capability Scorecard

Item	Level	Summary	Comp.	Opt.	Success Rating	Weight Cat.	Max. Score
R4-001	4	Change and configuration	Y/N	N	0–5	3	15
R4-002	4	Contingency plans	Y/N	N	0–5	3	15
R4-003	4	Emergency action plan	Y/N	N	0–5	3	15
R4-004	4	Crisis management team	Y/N	N	0–5	3	15
R4-005	4	Design assurance	Y/N	N	0–5	2	10
R4-006	4	Infrastructure transformation	Y/N	N	0–5	3	15
R4-007	4	Backup and tape management	Y/N	N	0–5	3	15
M-016	4	Change communication	Y/N	N	0–5	2	10
M-017	4	Documentation	Y/N	N	0–5	2	10
M-018	4	Project governance	Y/N	N	0–5	2	10
S1-009	4	Review security procedures	Y/N	N	0–5	2	10
S1-010	4	Resource vetting	Y/N	N	0–5	2	10
S1-011	4	Deployment management	Y/N	N	0–5	2	10
S1-012	4	User access control	Y/N	N	0–5	2	10
S1-013	4	Data center security	Y/N	N	0–5	3	15
S2-007	4	Review safety procedures	Y/N	N	0–5	2	10
S2-008	4	Data center safety	Y/N	N	0–5	3	15
Total:							210

Level 5

Capability scoring at this level may be recorded utilizing Table 11.6.

Table 11.6
Level 5 Capability Scorecard

Item	Level	Summary	Comp.	Opt.	Success Rating	Weight Cat.	Max. Score
R5-001	5	Design review	Y/N	N	0-5	3	15
R5-002	5	Server consolidation	Y/N	Y	0-5	3	15
R5-003	5	Storage consolidation	Y/N	Y	0-5	3	15
R5-004	5	Enterprise application integration	Y/N	Y	0-5	3	15
R5-005	5	Common operating environment	Y/N	Y	0-5	3	15
R5-006	5	Distributed systems	Y/N	Y	0-5	3	15
R5-007	5	Infrastructure renewal	Y/N	N	0-5	3	15
R5-008	5	Harmonize environments	Y/N	N	0-5	2	10
M-019	5	Manage changes	Y/N	N	0-5	1	5
M-020	5	Communicate changes	Y/N	N	0-5	1	5
M-021	5	Project governance	Y/N	N	0-5	2	10
M-022	5	Project quality review	Y/N	N	0-5	2	10
S1-014	5	Server hardening	Y/N	N	0-5	2	10
S1-015	5	Secure communication protocol	Y/N	N	0-5	2	10
S1-016	5	Remove shareware	Y/N	N	0 5	2	10
S2-009	5	Implement safety infrastructure	Y/N	N	0-5	3	15
S2-010	5	Locate safety equipment	Y/N	N	0-5	1	5
S2-011	5	Maintenance procedures	Y/N	N	0-5	1	5
Total:							200

Level 6

Capability scoring at this level may be recorded utilizing Table 11.7.

Level 7

Capability scoring at this level may be recorded utilizing Table 11.8.

Table 11.7
Level 6 Capability Scorecard

Item	Level	Summary	Comp.	Opt.	Success Rating	Weight Cat.	Max. Score
R6-001	6	Replication assessment	Y/N	N	0–5	2	10
R6-002	6	Data integrity	Y/N	N	0–5	3	15
R6-003	6	Data architecture review	Y/N	N	0–5	3	15
R6-004	6	Time server	Y/N	N	0–5	2	10
R6-005	6	Distributed systems fail-over	Y/N	N	0–5	3	15
R6-006	6	Consolidation in HA	Y/N	Y	0–5	2	10
R6-007	6	Environment location/strategy	Y/N	N	0–5	3	15
R6-008	6	Implement resilience capabilities	Y/N	N	0–5	3	15
R6-009	6	Resource resilience	Y/N	N	0–5	3	15
R6-010	6	Work management system	Y/N	N	0–5	2	10
R6-011	6	Workforce requirements	Y/N	Y	0–5	2	10
R6-012	6	Process review	Y/N	N	0–5	2	10
M-023	6	Manage changes	Y/N	N	0–5	1	5
M-024	6	Communication	Y/N	N	0–5	2	10
M-025	6	Project governance	Y/N	N	0–5	2	10
S1-017	6	Resource planning	Y/N	N	0–5	3	15
S1-018	6	Backup security agreements	Y/N	Y	0–5	1	5
S1-019	6	Security regression test	Y/N	N	0–5	2	10
S2-012	6	Implement backup equipment	Y/N	N	0–5	3	15
S2-013	6	Backup safety agreements	Y/N	Y	0–5	1	5
Total:							225

Table 11.8
Level 7 Capability Scorecard

Item	Level	Summary	Comp.	Opt.	Success Rating	Weight Cat.	Max. Score
R7-001	7	Recovery profile	Y/N	N	0–5	3	15
R7-002	7	Performance objectives	Y/N	N	0–5	2	10
R7-003	7	Dependency mapping	Y/N	N	0–5	2	10
R7-004	7	Interface harnessing	Y/N	N	0–5	2	10
R7-005	7	Disaster recovery plan(s)	Y/N	N	0–5	3	15
R7-006	7	Revise contingency plans	Y/N	N	0–5	2	10
R7-007	7	Event classification and declaration	Y/N	N	0–5	3	15
R7-008	7	Update EAP and CMT documents	Y/N	N	0–5	2	10
R7-009	7	Environment strategy	Y/N	N	0–5	2	10
R7-010	7	Data replication process	Y/N	N	0–5	2	10
R7-011	7	Promotional model	Y/N	N	0–5	2	10
R7-012	7	Software licensing	Y/N	N	0–5	2	10
R7-013	7	Recovery center location	Y/N	N	0–5	3	15
R7-014	7	Workplace recovery	Y/N	N	0–5	3	15
R7-015	7	Implementation strategy	Y/N	N	0–5	2	10
R7-016	7	Testing strategy	Y/N	N	0–5	2	10
R7-017	7	Implement recovery capabilities	Y/N	N	0–5	3	15
R7-018	7	Test capability	Y/N	N	0–5	3	15
H7-019	7	Final process updates	Y/N	N	0–5	2	10
M-026	7	Manage changes	Y/N	N	0–5	1	5
M-027	7	Communication	Y/N	N	0–5	2	10
M-028	7	Project(s) close-out and audit	Y/N	N	0–5	2	10
M-029	7	Debriefing process	Y/N	N	0–5	1	5
S1-020	7	Audit facilities (security)	Y/N	N	0–5	2	10
S1-021	7	Review transport (security)	Y/N	N	0–5	1	5
S1-022	7	Security planning in disaster	Y/N	N	0–5	1	5
S2-014	7	Audit facilities (safety)	Y/N	N	0–5	2	10
S2-015	7	Review transport (safety)	Y/N	N	0–5	1	5
S2-016	7	Key personnel safety	Y/N	N	0–5	2	10
S2-017	7	Resource location in disaster	Y/N	N	0–5	1	5
S2-018	7	Post-event care	Y/N	N	0–5	1	5
Total:							310

11.2   Capability Scoring

You will see from the rating tables above that each task has been assigned a weighting category. There are three categories for weighting in the continuity assurance model, numbered 1 to 3. The weightings given are intended to represent not only the relative importance of the task to the success of your capabilities, but in some cases, also the relative effort required to perform the task. Some tasks are relatively simple and involve initiating a review or producing a report. Others, however, can involve the implementation of significant change to your organization and its infrastructure. Indeed, many of the tasks represented in these tables are projects in their own right.

This system of rating may seem like an oversimplification to some, and in fact it is. It is intended to be so, since being too granular in assessment for something as complex as this would lead to an unmanageable rating system, which would be of little practical use. Fortunately, most things do come down to the simple fact of: did it work or didn't it, a binary answer. I have given a little more granularity with the 0 to 5 achievement rating. I do not believe that there is any advantage in making this any more granular.

We rate success at each level of the model in terms of a percentage, that is, your total achievement rating over the possible maximum at that level. The benchmark for success at any level, and progression to the next, is 50%.

If you have scored a rating of three or better on the majority of tasks then you should achieve this easily. You could score enough on some tasks that you could get away without doing some other tasks in terms of score, but remember that all mandatory tasks need to be addressed. If you have a score greater than 50%, but have mandatory tasks that have not begun, then you will remain at the level you are at until you address these tasks.

Scoring should be done by a panel chaired by the continuity assurance group leader, whether this is a manager or consultant within your organization. You can either have each member score separately, and take an average, or you can all agree on the ratings given for each task.

11.3   Maturity Progression

Continuity assurance is a maturity model and in rating your performance at each level we are determining success in meeting the criteria at that level, and your ability to progress to the next level of maturity. Remember that the model must be followed in sequence from one level to the next. Note that the tasks within each level do not necessarily need to be done in sequence within that level, you should use common sense when planning the order in which tasks are performed. If you get the sequence wrong it is sure to show up in your achievement ratings.

You will have noted that it is possible to progress to the next level in the model with some tasks from the level you are at not complete. This has been deliberately designed into the model for reasons of pragmatism.

If you are adopting the CAM and already have some business continuity capabilities in place, then it is possible that you have already achieved a level of capability at, say, level 7, before you have started level 1. This is expected; however, it is unlikely that you have achieved your full potential with this capability without having achieved the preceding capabilities in the model.

You should continue to work through the model and revise your capabilities where a gap presents itself. If you have only achieved one or two criteria at level 7 this does not rate your organization with continuity assurance capability at level 7. Of course if you go through the model and find that you already have met all criteria then, as long as you can substantiate this, you should be given the corresponding overall rating.

We have talked about tactical and strategic solutions to problems. The methodology is largely strategic in nature but this does not preclude you developing a tactical capability at later levels of the model than where you are currently rated. However, these tactical solutions should be revisited when you arrive at the relevant level of maturity and possibly reworked with the knowledge that you have obtained at previous levels of the model. Again, you could have a tactical capability at level 7 but still only be rated at level 1, or not at all.

The continuity assurance methodology incorporates, as a core principle, continuous improvement. Just because you have attained a satisfactory score at level 1, and have moved to level 2, it does not necessarily mean that activity ceases on level 1 activities. Additionally, once you reach level 7 you are required to continuously loop through levels 1 to 7, improving or maintaining your capabilities. If at any time you reassess at, say level 2, and you do not meet the minimum requirements at that level then you will be relegated back to a level 1 capability rating. In this way we make sure that continuity assurance keeps step with ongoing changes within your organization.

It should be said that we do not rate an organization in terms of overall score across all levels of the model. Your organization's continuity assurance capability should be communicated simply as level 1, level 2, and so on.

11.4 Audit

If you wish to utilize or communicate your achievement for commercial advantage, insurance reduction, or any other such purpose, you should ensure that your rating process is audited by an external agency. You may wish to have an external agency that has the necessary expertise perform the rating process for you, or at least concur with your results. This should give additional credibility to your achievement.

12

Quality Assurance

The assurance of quality is a key component of the management functions within the continuity assurance methodology. This chapter gives a précis of quality assurance principles that should be applied when implementing continuity assurance.

12.1 Fundamental Concepts

Quality assurance is a function that should be utilized across all processes. It is a measure of the adherence to defined process. In defining and adhering to the processes used to produce products, services, or project deliverables, you ensure a measure of quality of these outputs.

In our quest for superior quality we implement our own processes to monitor quality across any activities in which we partake. These quality processes are also intended to maintain a measure of business, or project, control, and ensure product integrity.

Quality management processes should be defined by all functional groups within the continuity assurance organization. They should be both embedded within the group's operational processes, and also act across them. These processes should be reviewed and agreed with the organization's quality assurance leader who will ensure that they are adhered to in use, and updated when required. The quality assurance group will further audit these quality processes on a regular basis. In this way we construct a nested process monitoring capability, where functional teams monitor their own quality, and the quality assurance team monitors them. It is too much to ask that one team is responsible for quality across the organization. Utilizing this approach we devolve some quality assurance responsibilities to the frontline of business operation.

Some of the fundamental principles of quality assurance are:

- Expectations of management, shareholders, stakeholders, and customers must be known and provided for.

- Every activity within the organization must be known and referenced as part of a defined and documented process.

- Quality should be assured across the process in addition to being controlled in the inspection of resulting products.

- Quality should be culturally endemic to the organization.

- Quality is an asymptotic process in that we continually strive to better it without ever reaching a maximum. Striving for quality is a never-ending process.

12.2   Framework

In terms of application to continuity assurance activities quality is assured across each level of the model and controlled at the boundary between each level. Quality, to some degree, is built into any methodology as a methodology provides the basis of an overarching standard that is being followed in the pursuit of betterment.

If we review the inner rings of our methodology's circular representation, we find that the diagram represents quality as a uniform layer, acting beneath and within the intervals in between the delivery levels of the model (see Figure 12.1).

Just as continuity assurance is itself an iterative and continuously improving system, so to are the quality assurance principles that bind it.

This may be represented in a more conventional linear fashion shown in Figure 12.2.

In this way quality is controlled at the boundary between each level and assured across all levels of the model.

12.3   Purpose and Strategy

It is the purpose of quality assurance to ensure that all activities are performed in a controlled fashion; that the outputs at each stage of a given process are measurable, and such measurements fall within a range set out within quality metrics.

The quality assurance function should address the auditability of all processes, and accuracy of both inputs and outputs to these processes. Quality

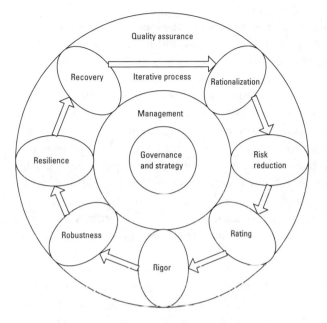

Figure 12.1 Quality assurance in the continuity assurance framework.

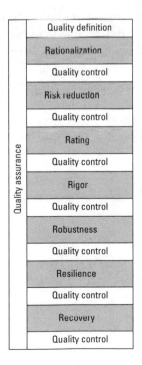

Figure 12.2 The quality method.

assurance also has a role to play in such "soft" functions as communication and knowledge transfer. It should assure and measure both the thoroughness of the undertaking and the effectiveness of the overall process in terms of results.

Quality assurance is itself an exercise in risk management. Adherence to quality assurance principles both optimizes the effectiveness of process and minimizes the probability of error. In ensuring that all work is performed as part of a discrete process, we help to guarantee that no activity is overlooked or omitted. The quality assurance process will itself highlight potential continuity problems within the organization by highlighting gaps or inefficiencies in processes.

Figure 12.3 shows quality assurance as being positioned around the operational process, and present at the input and output interfaces. This is in addition to the presence of quality assurance within the process itself, that is, integrated into the process. In this way quality is present both internal and external to each and every process.

This is true quality assurance in operation. You will hear people speak of quality control, and I have also used this term carefully at times in this book. It is important to understand the difference between control and assurance. Quality assurance is derived from the principles of total quality management (TQM), which was originally developed for use in manufacturing engineering to address issues with traditional quality control.

If we consider a manufacturing example, say a production line of widgets, quality control would involve the inspection of the widget at the end of production. Even though the production error may have occurred at an early stage of the process the fault would not be detected until the end of the production run. The faulty widget would then need to be either disposed of or reworked through every stage of the process again. This is clearly time consuming and inherently inefficient. If we rework Figure 12.3 to represent quality control we would arrive at a linear process that would look something like Figure 12.4.

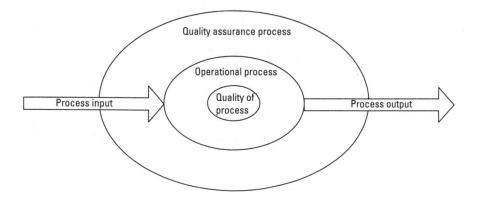

Figure 12.3 Quality assurance process model.

Figure 12.4 Quality control process model.

In attempting to make this quality process more efficient engineers came up with the idea of implementing quality processes at each stage of production. Furthermore, they looked at the actual activities at each stage and tried to improve the way in which the activity was being conducted to attempt to reduce the number of faults. This new process was termed quality assurance. Quality assurance, unlike traditional control, was both an active and interactive process of reducing risk of error through better process management. Engineers found that faults were greatly minimized in number; and, where they did occur, they were detected earlier in the production process and could be either reworked through fewer steps or disposed of immediately, sparing wastage of processing capacity on an already faulty widget.

I should say that the evolution from quality control to assurance is not dissimilar to the evolution that is represented in this book from traditional BCM to continuity assurance.

12.4 Quality Balance

I have talked about the balance to be struck between continuity assurance and sensibility; this balance point applies similarly to quality assurance.

You will find that if your processes are too rigorous you may stifle productivity and ultimately reduce quality of product. It is important to be able to strike a balance between process rigor and effectiveness of process. Remember that our main goal is the continuity of your business. For continuity to be achieved, processes must be functional and fit for purpose.

The principle aims of quality assurance are to:

1. Enforce policy;
2. Ensure clarity of activities;
3. Effect auditability of process;
4. Support traceability of components and products;
5. Expose issues with process or product;

6. Manage resource effort and responsibility.

Through achievement in each of these six facets we effect assurance of quality across the organization.

Again, there is a point at which a process can be unnecessarily thorough. When I say that processes should be fit for purpose, I include within this criteria the measure of rigor or intensity associated with the process. If processes are too detailed or there are too many processes to follow people can become bogged down in the processes and lose sight of the overriding business objectives at hand. In this way a process that is not fit for purpose can be counterproductive to both quality and continuity alike.

The graph in Figure 12.5 represents the "normal" distribution of quality against intensity of process.

The tip of the curve represents the balance point between quality and process intensity, or control and delivery. We call this point the optimum quality point or OQP. This is the point where quality is optimized by a sensible and pragmatic level of process granularity and control.

12.5 Approach

The implementation of quality assurance principles can be broken down into the following high-level areas:

1. Process management;
2. Quality culture;

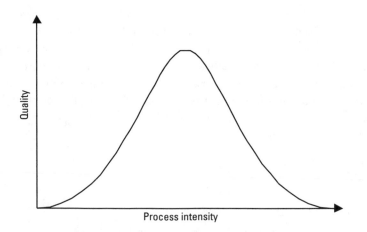

Figure 12.5 Quality distribution and optimization.

3. Checkpoint assessment.

These are all self explanatory. Both qualitative and quantitative assessment of effectiveness should be applied to each of these areas.

Quality assurance should be applied equally to both project and business-as-usual activities. There are a number of performance indicators within continuity assurance that entail conducting quality checkpoint reviews or review of overall process quality. These are detailed within the levels of the continuity assurance model.

There are many ways to implement and measure quality and I do not intend to go into them all, as the existing body of knowledge on this subject should be sufficient. I personally find the following adaptation of the earned value model (see Figure 12.6) for project management quite useful in quality assurance terms, for projects at least. Again, this is not a book on project management so I will not go into a great deal of detail.

Essentially, over the course of your project you plot out the following curves on a graph of days of effort versus elapsed time. You could use cost instead of days as a measure of effort but cost information is often more time consuming to compile.

The various traces on the chart are as follows:

1. *Baseline plan time.* This is the initial agreed version of the master project plan.

2. *Total time* is the time billed or recorded by resources on your time-reporting system, timesheets, or the like.

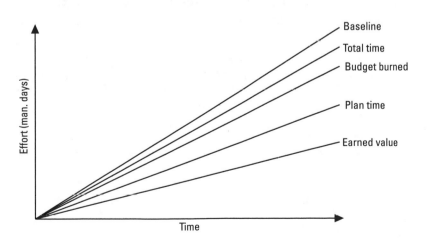

Figure 12.6 Adapted earned value chart.

3. *Budget burned,* or rather time burned. This is time recorded by resources against tasks allocated within your work management system.

4. *Plan time* represents the current floating plan, which should include new tasks that represent faults, changes, and issues raised throughout the course of the project.

5. *Earned value.* The concept of earned value is such that you are only given credit for a task once it is completed. For example, if a task on the plan is allocated 4 days then you may burn 5 on the task and not complete it. Once completed, even if you spend 10 days on it, you will only get credit for 4 days. For faults, issues, and so forth contained in your work management system, it is useful to utilize a two-stage earned value, giving 80% of value for resolved tasks and 100% for closed tasks. Note that if a task is reopened you lose all value previously gained.

So how does this relate to quality assurance? Well, we can now apply some statistical analysis to the data represented on the chart. You should find the following very useful indicators both quantitatively and visually.

- *Slippage* can be measured as the difference between plan time and earned value.
- *Quality* can be measured as the difference between the baseline and plan time, as the plan time includes the impact of all recorded faults and problems.
- *Efficiency* is the difference between earned value and budget burned.
- *Productivity* is the difference between budget burned and total time.

13

Outsourcing and Commercial Management

Many of the tasks that need to be undertaken in order to implement continuity assurance are quite substantial projects or operational initiatives in their own right. At many points you will be faced with the option of outsourcing, whether it be to run a project or to provide a recovery facility.

13.1 Insource or Outsource

In terms of projects, you will remember from our discussions around the rationalization of your organization that it should be possible to run most continuity assurance projects utilizing the continuity assurance team that you have put in place; perhaps assisted only by a contracted project manager and some additional specialist technical resources, depending on the nature of the project. However, there may be cause to completely outsource some types of projects.

In terms of project outsourcing I would recommend that you selected a well-known and respected systems integrator. After all, one good reason to outsource is to disperse risk, risk of failure that is. The projects that will need to be undertaken as part of continuity assurance are complex and difficult. If you do not feel comfortable that you have the skills to manage these yourself then you are best to leave it to someone that does. You should pick an organization that has the financial robustness necessary to withstand any problems or issues that may be discovered throughout the life of the project. Choose an organization that has a reputation to uphold and is willing, and more importantly able, to absorb a measure of movement in the projects activities. That said, be aware that if you push them too hard on timescales and scope while being inflexible with remuneration you may diminish their ability to deliver at all. If you wish a subcontractor to be flexible you must also be flexible. Playing games with commercial arrangements and trying to bleed extra services from your subcontractor is not the way to ensure a healthy outcome to your project. Both you and the subcontractor must work as one team to make the project a success. You must use a measure of commercial empathy; understand their commercials as well as understanding yours. Do not put them in a position where you know they are losing money and expect them to perform at their best. I have said before that there are situations where it does not matter what is written in a contract, you cannot contract a person's helpfulness, or motivation, or pride in their work.

Be aware that many organizations you engage will simply go out to the contract market and buy themselves the resources they need to do the work for you, under their name, and at sometimes significant markup. You will have no idea who works for who in your project team. Note that this practice is certainly not limited to small organizations, in fact my experience is that it occurs more often in larger organizations. Smaller companies cannot afford to bankroll contractor's rates. Why does this happen? Well most companies cannot afford to have these sorts of expensive resources on retainer, at least in the numbers that you need to perform a large-scale project. However, there is nothing stopping you from going to the contract market and obtaining the same resources for yourself. Unfortunately, the responsibility for delivery will rest with you in this case.

Understand what value the company you are considering brings to the table. Sometimes it may simply be their ability to manage the contractors and the commercials, or it could be their commercial leverage with equipment suppliers or similar. Do not be blinded by the branding of large institutions. Understand their business and you will be in a position to make the best use of them. In many cases you may simply wish to utilize the effect of their branding to have a consultant convey a message to your management that will look better coming from an external company than yourself. You may utilize the consultancy simply to give credence to your ideas or methods. This is quite legitimate, and is unfortunately a fact of life in consulting and management.

In any outsourcing situation you need to have a strong governance system. Be conscious of the overhead of this system in terms of cost. Often the cost of governance can outweigh the financial benefits of outsourcing in the first place. If you have to put together a team of people to keep another team of people on track then it starts to get a little ridiculous. I have even experienced the situation where a company had one consultancy doing a project and another consultancy governing them. This is not an efficient way to conduct your business.

I feel the need to say a word or two about consortiums. A consortium, in project or program terms, is where a number of companies are brought together to undertake a large-scale project, usually because no one company has the expertise to deliver the project in its entirety. These can be complicated to control, especially in commercial terms. If you are going to do this it is better to contract one consultancy to lead the consortium. This consultancy will coordinate the project and the other companies. Commercials can either be passed through the prime consultancy or negotiated directly, with the prime consultancy having operational control only. Again, strong governance is required.

Be aware of the potential security issues regarding outsourcing facilities, operations, or projects. Remember that the personnel that your subcontractor uses should be subject to the same security vetting process that you use for your own staff. You could try to control this contractually and this will cover you in terms of liability should something go wrong, but does it help you in terms of preventing the problem?

Be careful that when deciding whether to insource or outsource that you add up the real costs involved. Do not be influenced too much by the perceived simplicity of outsourcing, often there are many hidden overheads and risks.

One of the benefits of the continuity assurance organization, if implemented correctly, is that you should need to outsource only minimal numbers of activities. Remember that continuity assurance is a process of continuous improvement. It is not a capability that can be implemented in its entirety as a project and then handed over to a team to run. The resource required to run the processes in a business-as-usual setting is really no less than the number and skill set of those required to implement the initial capabilities. In any event, if you are just starting out in business continuity within your organization it is likely to take some time to get to a respectable capability level, so best to retain the staff that can make it work for you.

13.2 Recovery Site Providers

One principle area where many utilize external services is that of provisioning a recovery facility, be it for systems recovery or workplace recovery.

In many cases you will find it is the same companies that provide joint systems and workplace recovery facilities. There are quite a number of these organizations throughout the world. Indeed a handful of these companies have a network of centers across the globe. In some cases though, you may find that any company with some spare data center capacity that they want to utilize will call themselves a recovery site provider. Watch out for this.

The biggest issue that you will face with these providers is that they will typically provide services on a share basis. This means that a number of organizations subscribe to utilize the infrastructure and workspace in the event of disaster affecting their particular company. The recovery provider ensures that they can meet the equipment requirements of their customer within their infrastructure, and then charges a fee for the option of use in a disaster situation, and a subsequent fee should you actually need to use it. Remember that you are not renting the facility for all time; you are merely purchasing an option to use the facility in the event of disaster.

The obvious problem with this type of subscriber-based facility is the fact that the infrastructure is shared. What happens when two subscribers suffer a disaster at the same time? The answer is that the system basically works on a first-come first-serve basis. Whoever declares the disaster first will have access to the infrastructure that they require. If some of the infrastructure that is part of your particular configuration is still available then you may be able to use it, but this will probably be of little use to you without the rest of the infrastructure you need. Remember that if you suffer a regional disaster or a continuity-impacting event that affects a number of companies at the same time then there could be a rush on the recovery center.

The other key issue with shared sites is that they do not generally keep an active image of your systems on their infrastructure. The hardware will be clean and ready to be used by any subscriber. This clashes with the capabilities we require for some systems under the continuity assurance methodology. The systems that require a cold standby capability will be the only systems that lend themselves to this type of shared facility. This is not an unusual situation and most facility providers will provide, at a cost, rack space for you to install your own dedicated servers. The inference is, of course, that you will need a dedicated system for any capability higher than cold standby. You will typically also need your own dedicated wide-area network routers into the facility, which you will have to purchase. Clearly if a large percentage of your systems require better than cold standby capability then you should probably look at other alternatives than a shared facility. You could have your own facility for hot or better systems and a shared facility for cold, but this is unlikely to be cost effective. You would need two sets of routers for a start. This sort of mixed solution is not recommended. It overcomplicates something that is complicated enough to manage in the first place. Remember that the simplest solution is most often the lowest risk solution.

Most recovery facilities companies will do both systems and workplace recovery, as I have already said. You will notice that most do one better than the other, there are few that do both well. Make sure you look around and do your homework on what is available in your required location. Depending on the size of your operation there are companies that will upgrade their facilities in order to get your business. In the case of a shared facility it is sometimes better to go with a smaller organization. For a start, they are likely to have less subscribers so your risk of not being able to use the facility in a disaster will be lower. They are also likely to be more flexible in terms of contract and adjusting their business to support your particular needs.

Most shared facilities work on the basis of a standard contract to all subscribers. They are quite inflexible. In fact their whole business model relies on this inflexibility. The problem is that if they alter a contract for one subscriber it may give that subscriber some level of preference over the infrastructure. This immediately diminishes the value of everyone else's contract. This would eventually result in the demise of their business. For this reason the commercials are generally black and white, at least with the larger players in this business. You either take the standard contract for a shared service or you take a dedicated service. A dedicated service is basically you purchasing, or renting, all of the equipment you require and locating it in their recovery facility.

All facility providers will provide a level of technical and logistical support. Again the level of service they provide varies greatly so be sure of what you need and what you are getting.

Be particularly attentive to the level of security provided. There is of course an inherent security risk in using an external provider in the first place. You need to be sure that the facility provider meets all of your security requirements and satisfies the security requirements under the continuity assurance methodology.

The facility you choose for recovery services will be your recovery data center. It should meet the requirements of data centers under this methodology. From my experience there are very few that meet these standards completely, so be cautious. Make small compromises where required, but try to put a plan in place with the facility provider to upgrade facilities over the term of the contract. Any compromise will clearly affect your capability ratings but the methodology is designed to allow for such pragmatism. Continuity assurance is about continuous improvement.

A word on testing; most facility providers will make provision in their contract for a certain number of days of testing which is included in the standard terms. This test time will need to be booked some time in advance and you will need to fit in around other subscribers, in the case of a shared facility. Again, time is booked on a first-come first-served basis. It will normally be possible to book additional test days at a cost, provided the facility is available.

Note that if you are testing and another subscriber declares a disaster you will have to get out quickly. In such an event you will have the internal organizational cost of having to repeat your test at another time. Depending on your size, this cost can be quite large. It requires a great deal of organization and involvement from a large number of people across the organization to conduct a proper disaster recovery test. Take costs such as this into account when you are deciding whether to insource or outsource such facilities. Decreased control will often lead to increased risk and increased cost. That is, control is inversely proportional to risk and cost.

13.3   Request for Proposal (RFP)

It is worth mentioning a few words on writing RFPs. This is different to writing a tender document or request for tender (RFT). A tender is used when you know pretty much exactly what you want and just require a price on it, for example, 100 desktop computers. An RFP is used where you have an idea of what you want, in that you can stipulate some high-level requirements, but you want someone to give you a proposal for solving your problem.

In general, if you are insourcing then you will tender for equipment and the like, and if you are outsourcing you will ask for a proposal to address your requirements. In many cases companies will contract a consultancy to produce the RFP and possibly gather the initial requirements. This work could be commissioned based on the outcome of a previous, much simpler, RFP process to select the consultancy, or you could simply commission them on a time-and-materials basis. Of course you could do it yourself, which should be more than possible with the continuity assurance organization that you have put in place.

RFPs should be clear and simple, and written in plain English. Where possible, try to keep the length of the document to a minimum. Representing requirements in tabular form makes the job of the respondent simpler. Most companies spend a great deal of money on putting together specialized teams to respond to RFPs. The respondent company will assess the amount of time and effort needed to respond versus the potential remuneration and strategic advantage resulting from a successful bid. To give you an idea of the investment that these companies make, an RFP worth about $10 million in gross value will cost a company perhaps $200,000 to respond to. This is a great deal of money to spend without any guarantee of obtaining the work. In addition, there is the opportunity cost to bear if they are not the successful vendor. Companies only have a small number of people capable of doing this work. While they are responding to your RFP it is likely that they are not responding to someone else's. RFP responses involve a great deal of consulting work, which you get for free. Indeed, there are some companies that will put out RFPs just to get an idea of how to do the project themselves. They take the information obtained from the response and put their own team together to execute it. This is of course highly unethical but it does occur.

It is important in business to have a degree of empathy with your suppliers. It will help you to get the best and fairest deal for your organization and theirs. Projects should generally be run more like a partnership than a customer-supplier-type relationship, in operational terms at least. You need to understand that in most cases your respondent knows more about this topic than you do, otherwise you would not have asked them. Listen to what they have to say.

Make sure your RFP states what you are wanting in requirement terms, but do not give the solution, even if you have one in mind. It is best to keep the RFP open in terms of solution. If the vendor thinks you have something in mind they are likely to give you exactly that, because they are aware of the effort and risk involved in trying to change your opinion. You may be preventing a much better solution from being presented to you. Leave provision for the respondent to question your requirements to some degree. We have discussed earlier about wants and needs not necessarily being the same thing. You want the respondent to tell you what you need to make the project a real success. You may be asking for something that first requires another five things to be implemented before you will be ready to deal with it. It is akin to you putting out an RFP for a level 7 capability and me responding with all the things at levels 1 to 6 that you should do first. Make sure that you have a clause in the RFP that enables you to withdraw it and pursue independent negotiations if such a situation arises; that is, where you have clearly got it wrong.

Involve your lawyers at the earliest possible time and certainly before the RFP is compiled. The lawyers will need input into the RFP document and early involvement will ensure that draft contracts are available quickly, as the contract will largely be based on the wording in the RFP. Do not simply write the RFP and then pass it to your lawyers to check it. Remember our distinction between quality assurance and quality control. You need to integrate legal resource into the team putting together the RFP. This may well cost you more up front but, as with quality assurance, it will save you money in rework and time.

Ensure that a pricing template is enclosed with the RFP so that responses can be easily compared, apples to apples.

13.4 Running a Commercial Process

The typical commercial process for outsourcing projects or facilities should look something like this:

1. Produce RFP.
2. Determine RFP distribution.
3. Conduct vendor briefing, advising of imminent release of RFP.
4. Issue RFP.
5. Receive intent to respond.
6. Determine selection criteria.

7. Answer questions.

8. Receive proposals.

9. Assess proposals.

10. Shortlist vendors.

11. Vendor response briefings.

12. Inspect facilities (where relevant).

13. Determine successful vendor.

14. Issue letter of intent (LOI).

15. Contract negotiation.

16. Execute contracts.

Some of this can be varied to speed up the process. For example, it is possible to have the vendor conduct a briefing prior to them submitting their final proposal.

It is imperative, ethically speaking, that the selection criteria be decided before any responses are received. Note that these selection criteria can include provision for assessment of response information over and above what was specifically requested.

Timeframes are important and need to be set from the start of the process. Give the companies you select as much time to respond as you can. It should ensure that all companies are able to respond and should also increase the general quality of responses. Again, be open to new ideas and answer questions as honestly as possible.

When answering questions you should make both the question and answer immediately available to all respondents that have not yet been eliminated.

In some countries, or for some type of work, it is usual to ask for what is termed a bid bond. This is essentially a letter of credit issued by the respondent in your favor and is submitted prior to their proposal being accepted for consideration. This is essentially in recognition of the fact that you will be putting a lot of effort into assessing their response, and at the end of the day may require them to accept in order to gain value for your efforts. This bid bond is essentially used to gain their commitment to proceed should you select them. They can withdraw or decline but they lose the bond. You could cover this commitment contractually but remember that contracts are not negotiated until the end of the process. It is possible to bind a company to proceed in the RFP and subsequent response but this is not always easily enforceable, depending on your legal jurisdiction, and at any rate will cost you money to enforce.

In addition to the bid bond some organizations will ask for a performance bond, again in the form of a letter of credit in your favor should they fail to perform on the contract. These are all things to consider when planning out your commercial process, and to some degree governance as well.

13.5   Vendor Selection

As I have mentioned, you should have a clear rating mechanism with which to assess vendor responses. Your RFP should detail any areas where a level of legal acceptance of terms is required. You rating process should consider:

- Compliance to request;
- Format and quality of response;
- Completeness;
- Willingness to deliver;
- Ability to deliver;
- Pricing and hidden costs.

Assessment should consider the RFP response, presentation, and any questions and answers processed throughout the commercial process.

In some cases you will receive a response from one or more vendors teaming together. This should be assessed in terms of risks associated with such an approach. There should be one clear lead proposer that can be contracted. Remember that this may complicate the contract process, as they will attempt to back off the agreement with their subcontractors. This can lead to protracted negotiations.

You should attempt to rate the respondents both quantitatively and qualitatively, and convert results to a common quantitative format. Aspects of the assessment such as ability to deliver are critical. Some factors should be weighted higher or lower than others.

Ability to deliver should be based partly on your understanding of the respondent's commercial situation. If they are bidding something that you know is costing them more than they are selling it for then although they may be keen to deliver, as a loss leader, they may come to a point where they are not able. I cannot stress too much the importance of rating ability to deliver.

Make sure you read and understand the responses. Do not preclude a company on the basis that they have not answered as you may have expected, you may be eliminating the best solution. In reading the responses you may find that you have asked for the wrong thing or not been clear enough in what you require. If a vendor is pointing this out to you then it probably represents a great deal of integrity on their part. You may wish to abandon the RFP process and enter independent negotiations with this vendor.

In terms of assessing price, make sure that you are comparing apples to apples, and that you expose any hidden costs. Assess what you are getting for your money and whether a higher price is worthwhile. Many people immediately eliminate the highest and lowest price but this is an oversimplification. Assess responses on the basis of risk, after all these are continuity-assurance-related projects or services. My advice is to go with the lowest risk proposal but this is a judgment you will have to make based on the exact nature of the activity you are outsourcing, and the facts as they present themselves in the proposals.

When you are issuing an RFP for the provision of a subscriber-based recovery facility make sure you ask the respondent to enclose a copy of their standard contract with their response. Also ask the question as to how flexible they are on terms.

13.6   Negotiating Contracts

Contract negotiations can take longer than you may think. It is useful to keep in mind that length and complexity of negotiation is often not related in any way to the value of the contract being negotiated. It can take the same amount of money, time, and effort to negotiate a $1-million contract as it can for a $100-million contract.

Again, involve your lawyers and commercial managers in the process as soon as possible. These are the people that will have to conduct much of the negotiations and it is important that they know as much about the topic and history of the initiative as possible.

Contracts can generally be drafted immediately following completion of the RFP document. This means that you should have a complete draft contract to work with, as a starting point, at the beginning of negotiations. This will ensure that you are not starting from a blank sheet.

There are three things that I would like to highlight that are important when negotiating contracts for continuity assurance capabilities:

1. You are making provision for your operations in the event of a crisis or disaster. Pay particular attention to the force majeure clause in the contract.

2. Consequential damages will be near impossible to negotiate as these could far outweigh the value of the contract, considering the nature of the activity.

3. Where outsourcing facilities, you will almost certainly need to prove that you have software licenses for the software you are installing on the facility provider's equipment. Facility providers are unlikely to indemnify you in this matter.

When negotiating with a vendor that your organization has not previously used be aware that you may wish to utilize them for other future work. It is useful to allow provision in the agreements for subsequent unrelated orders or the like. This means that your agreement can be utilized to frame other agreements with this vendor in the future, and hence reduce future contract negotiation time, complexity, and cost.

13.7 Ethics

In any commercial process you should pay particular attention to probity and general ethics.

Conduct a fair process where all parties have an equal chance of success. Do not preselect a company that you wish to work with and go through the motions with other vendors just so that you can say you have executed a commercial process. This sort of whitewash is unethical and will not serve your organization in the long run. If such an activity is proved you could be open to legal ramifications. More likely you will become known for such practices and find it difficult to obtain participants in future RFP processes. It is not uncommon for companies to bow out of RFP processes because of suspected ethical concerns. As I mentioned, respondents will take a risk judgment on responding to your RFP based largely on their perceived chance of success.

Keep political agendas in check. You may feel compelled to select a vendor based on their ability to meet certain criteria or timescales that benefit you personally; perhaps in terms of bonus, for example, but may not be in the overall best interests of the organization. This I consider to also be unethical. It does not put all the respondents on a level playing field because the assumption of logic is not valid. They may propose something that is better for your organization but are unsuccessful because they are not in line with your own agenda.

It is useful to debrief unsuccessful vendors and give them some steerage on where they went wrong. You never know what other business you may wish to do with these companies in the future. Remember that they are likely to have invested significant time and money into responding to your RFP. I believe that they are entitled to some professional courtesy and encouragement.

14

Case Studies

The following two studies are included as examples of commonly executed approaches to BCM. You should be able to see the issues with the way these companies are doing things given the information you have absorbed from the previous chapters. You should also be able to see that not all of what is being done is wrong and that there are elements of the CAM that they are already using, albeit not always in the most efficient manner.

Note that the actual company names have been excluded from the text for fairly obvious reasons. To highlight potential gaps in continuity, or security, or general information regarding business methods would be to draw attention to potential vulnerabilities that could be exploited. To do this would contravene the methodology that I am promoting in this book. Hence these companies will remain safely anonymous.

14.1   Study 1: Telecommunications

Company X is a large telecommunications company. The company provides fixed line, mobile, data networks, and Internet services. The company is in a location that has an associated level of physical threat. Building regulations are poor as are general health and safety considerations. There are no particular natural hazards other than humidity at some locations and extreme heat. The company employs tens of thousands across multiple campuses and buildings. The great majority of IT systems are located in a single data center. Restoration in the event of disaster would be a rebuild from tapes stored off site. Physical security is heavy but poorly coordinated. Recovery from an event that renders the primary data center unusable could take many months. The systems

187

architecture has evolved organically over decades of use and it would be difficult to procure the skills necessary to reassemble parts of the infrastructure. The architecture is segmented with many subcontracted companies responsible for different systems. The systems selected are driven largely by the proprietary nature of the different brands of switch used. Company X is heavily reliant on its subcontractors and suppliers for tasks including day-to-day management of the network and systems.

As I have said, telecommunications is an IT-based business so virtually nothing takes place in business process terms without some level of IT interaction. The telephony networks are not typically considered IT systems but the switches connect directly to IT systems for control, monitoring, and billing of calls. It is useful to understand at this point how a telecommunications company operates. Figure 14.1 illustrates by way of simplified representation.

Calls are made using the switch network. The initiating switch (where the call begins) generates a call data record (CDR) for each call. This CDR is stored as a single file in the memory of the switch. The switch may check with other systems to ensure that the user is authorized to make the call. For example, for GSM prepaid calls the switch must check that the user has enough credit available to initiate the call. The switch is periodically poled by the management systems and the CDRs are uploaded (pulled) from the switch. The process differs somewhat where data network switches are involved depending on the exact service that has been provided to the customer (e.g., leased line, ISDN, and so on). The CDRs are then processed. This involves matching the call to a customer account and pricing the call. Information about the call and the charge is included in the customer's next bill. Each telecommunications company is different in the way their network is assembled and the number and type of

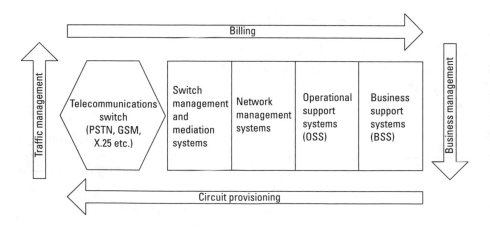

Figure 14.1 Simple telecommunications model.

different systems used in the business processes. The CDR will often pass through multiple systems before reaching the end billing system. In many organizations, including Company X, the CDR, or subsequent transformations of the CDR, will be transferred between systems using a simple batch transfer protocol to copy the file. Backups are stored both on-line for a period and to tape, for each system. A copy of the CDR may also be retained on the switch for a time. Hence the CDR is usually recoverable many times over. Indeed there is a large degree of data redundancy where the CDR is concerned.

It should be pointed out that switches are located throughout the areas of service in remote switching stations. Some of the switch management and mediation systems are also distributed. In the case of Company X, virtually all other systems are located in the central data center. Some have a level of resilience capability typically with the standby unit sitting alongside, or in the same rack, as the primary unit.

Company X identified the obvious vulnerability associated with having the majority of their systems colocated within the same data center. They established a business continuity team within the IT security department. This team was comprised initially of a small number of people with little or no previous experience of business continuity. This later grew to a significant number of people with little or no knowledge of business continuity. Many were transferred from other areas of the IT organization. Some were very astute and capable and knew the operation of the company well. Some were not. After several failed attempts at analyzing their holistic business needs around continuity Company X produced an RFP and distributed it to relevant consultancies, comprised of those with previous dealings with Company X and other leading companies providing services in this field.

The RFP was ambiguous in many respects but clearly focused on disaster recovery for IT systems. A consultancy was selected for an initial project to determine disaster recovery strategy. The same consultancy later went on to implement specific disaster recovery capabilities. Note that there was no, more generalized, business continuity department within the organization and no detailed business continuity work had previously been successfully completed across the organization. The first part of the project was for the consultancy to conduct a BIA and risk assessment with a view to determining the critical IT systems that should be provided for in a disaster recovery site. The consultancy wanted specific scope as the project was bid at a fixed price, so it was decided to ring fence the primary data center and limit provisioning of a disaster recovery capability only to those systems identified from the BIA as critical and that were located in the primary data center.

It is extremely difficult to perform business continuity projects based on a fixed price. It is impossible to determine the extent of problems that may arise from analysis. In confining the consultancy to a fixed-price bid Company X was forced to define a level of scope before appropriate analysis had been done. This also meant that the risk assessment was not done across the company but was only a risk assessment for the data center. Company X also entered the initiative right from the RFP stage with preconceptions of what the results would be. For example, they had already decided that their strategy would include a second, large, data center. In fact, as the consultancy later discovered, they had already selected a location and commissioned a company to build it.

These decisions would seem to have been made by executives with company-political agendas. For some the disaster recovery exercise was a rouse designed to procure funds for the expansion of their empires. The view seemed to be that of creating a secondary production facility under the guise of use for disaster recovery purposes. It should be noted that the newly appointed business continuity director had no part to play in these decisions, as he was several levels of authority below the decision makers in the organization.

The consultancy supplied a project team that was largely business focused or at least was significantly weighted towards the business. That is, the IT skills of many of the team members was inappropriate for the level of technical analysis they were expected to perform, given that this exercise, practically speaking, was one of IT recovery as opposed to holistic BCM.

The consultancy began work on the BIA by interviewing a plethora of people at varying levels within the organization, focusing on the business staff. Much of the information was opinionated and contradictory with each person interviewed placing disproportionate emphasis on the criticality of the processes that they were a part of, but this is to some degree normal. When the analysts attempted to drill down to the systems that supported the key business processes they encountered problems. Because of the segmentation of the organization and the fact that it was very top heavy, with multiple management layers and most of the "work" done by external companies or contractors, the exact systems in use for various processes became surprisingly difficult to identify. This was compounded by the extremely high rate of change of the technical architecture. Many systems were being phased out and being consolidated into new combined function systems and so on. There were also terminology problems with one system being known by several different names, depending on the exact function that was being carried out by a given business team. Add to all this the fact that there were other significant projects underway for storage consolidation, server consolidation, and EAI, and that all these teams had analysts out asking many of the same types of questions to the same people. This resulted in many members of staff being bombarded with requests from about four different consultancies at the same time. This led to problems with cooperation. The root cause of this problem was clear lack of coordination by the company's project management office. If the business continuity department was in the correct position within the organization this could have been avoided by enforcement of crossimpact assessment between the projects and then sequencing of the projects in some logical order. The analysis undertaken as part of the BIA and risk assessment could have been filtered down to the other project teams. Other projects could have fallen in line with the agreed business continuity strategy, which indeed would have recommended such initiatives under continuity assurance. Again, because of the segmentation of the organization these other projects were outside of the control of the business continuity director's immediate chain of command and the consultancy, when it came to capability creation, was instructed to work with these other teams and resolve any specific issues as and when they presented themselves. This of course proved problematic.

For example, when presenting design options for discussion some would lean towards an option that relied on server consolidation, because they may have had some involvement in this project. In such cases budgets and responsibilities can become clouded. If you propose consolidation as part of your business continuity initiatives then is it your responsibility to deliver it, or that of a team focused on consolidation. Do you take on the task or create an inter-project dependency. I know it seems a nonsensical situation but this is the sort of problem that eventuates from lack of corporate-wide strategy and project coordination. Remember that the consultancy is working on a fixed-price contract and any delays will cost them money. In the consulting business the saying "time is money" is never truer. So do they wait for the other project to come into synch with them or take on some of the tasks to help themselves out? This is an inherent problem when executing multiple projects with multiple vendors. Needless to say that Company X's vendor management left a little to be desired.

Going back to the question of strategy, the location of the recovery data center, which by this stage was being built, became a problem. The distance between the primary site and the selected recovery site was over 1,000 km (or 600 miles). This meant that synchronous real-time replication was not technically feasible, due to network latency. However, the BIA had revealed that this was the business requirement for at least one system. This is not to mention the obvious problem of relocating key technical staff across these distances, which incidentally was also outside the scope of the project and was hence largely ignored. Without going into the arduous detail here the result was the plan to later construct a third data center, to be used as a high-availability site. This all adds to continuity capability but for Company X this could have probably been avoided.

Remember that we are dealing with a telecommunications company where most of the interfaces between systems are batch transfer unlike a bank, for example, where most transactions are real time. Additionally, remember that even if the data center was wiped out completely the switch network would continue to operate. Most people would be able to continue making calls and the CDR would be stored on the switch. Only provisioning of new lines and billing would be stopped. Even if it took a week to get the systems back on-line in a recovery facility the CDR would still be collected and the customer eventually billed for the service. It is more than likely that the failure of two or more core switches on the network at the same time would cause more disruption to the business that the entire data center being wiped out. For this business continuity initiative the recovery of switching equipment and associated processes was firmly out of scope, due to the initial path and resulting sequence of events that Company X followed.

Another big problem that Company X had was with their technical management processes such as change management, configuration management, and general systems responsibility allocation. They failed to understand just how big a change they were embarking upon. Running multiple mirrored systems across multiple sites is not trivial to manage. It is a fundamental change to the core infrastructure of the organization and requires for success a re-engineering of all support processes. Not only did Company X not fully appreciate this, their history of success with such management processes in a single data center was less than perfect. They had underestimated the size of the task at hand and had not appreciated how fundamentally the implementation could change their organization. They were unprepared. It only follows that they had also underestimated how much management commitment, money, and time it would take to achieve success.

This is a prime example of the haphazard way in which many organizations approach business continuity. It has been said by many who have been through such exercises that good business continuity is achieved through a process of iterative improvement. The size of the task is often so big to grasp that one simply begins with the most obvious problem and proceeds to refine and consolidate their approach through continuous analysis and periodic capability testing. Indeed this is the approach that many organizations take. It is of merit that they are at least addressing problems and working towards better continuity. It is also disappointing that they waste so much time and money on methods based on little more than trial and error.

I do not feel comfortable disclosing exact figures even in an example where the company is not named but suffice to say that Company X has so far spent in excess of $20 million on completing this first stage of disaster recovery capability. They have taken a step in the right direction. Spending this sort of money you cannot doubt their resolve but they still have a long way to come. Following the CAM may have got them there faster and possibly more cost effectively.

As you will already know, disaster recovery is the last step in continuity assurance. It is acceptable to put in place some tactical capability for recovery as early as possible but a strategic, fully integrated, and successful disaster recovery capability can only be achieved by moving sequentially through the preceding six levels of continuity assurance. Note in particular that Company X has done nothing to attempt to reduce risks or resolve vulnerabilities or to align its organization for better continuity. They will ultimately have to move through all parts of the continuity assurance model at some time, perhaps multiple times, before their underlying business continuity objectives will be fully met.

14.2   Study 2: Mining and Energy

Company Y is a multinational mining and energy corporation. The company has a significant presence in the United States, United Kingdom, South America, South Africa, and Australia where it operates mines and processing facilities for raw materials. Because of the inherent dangers associated with Company Y's business their health and safety processes and procedures are second to none that I have seen. Risk management in general was also well done with the exception of any specific business continuity management function within the organization. This company operates in some politically unstable locations so security is an issue and is reasonably well managed when compared to similar-sized companies in this sector. Products in some instances have substantial tangible value, which further advances the case for heavy security.

On the surface it seems that this business is not very reliant on technology but executives feel differently, listing e-mail as one of their core business enablers. Because of the geographically diverse nature of the organization e-mail is often relied on for all manner of business process. Most processes are not highly automated and are done largely manually with e-mail acting as the core interface enabler between processes. A failure of IT systems does not impact the heavy machinery and workers per say, but many staff members do receive their work instructions via e-mail. Systems such as payroll and the like are of course critical to the continuity of the business, as are links to commodities markets.

Perhaps this explains to some degree why the company board set off down a similar path to Company X when they began their first foray into BCM through the implementation of a level of disaster recovery capability for core IT systems. The company's CIO was set a personal KPI concerning the creation of a basic recovery capability initially for operations in the United States and Australia, with a view for possible later expansion into all corners of the organization. The timeframe given for attainment of the high-level goals was short and reflected a lack of understanding of the complexities of such an initiative on the part of senior management. The KPI and the timeframe drove the deliverables sometimes at the expense of quality and strategic sensibility. This is not quite as silly as it first sounds because the company did go from having virtually no protection from disaster affecting their IT systems, to having a basic working capability in a short space of time. This meant that they were covered with a somewhat tactical response capability. This gave them time to develop a more strategic solution. It was accepted by many that an amount of rework would be required to reach the optimum solution for the organization. Indeed, the tactical solution only covered the core centers of business for the organization and would need to be expanded both in geographical area and technical extent. The thing to watch out for in situations such as this is that all parties understand that such an initiative is the tip of the iceberg, not the end solution for the company.

Company Y has a much smaller IT footprint than Company X, being a largely non-IT-related enterprise. Most of the critical systems were concerned with general desktop tools, e-mail, Internet, Intranet, and financial management systems. Interfaces between systems were largely noncomplex and file transfer-based. Due to the relatively low-tech nature of these systems most were Wintel–platform based. Though not complex in themselves, the sheer size of the organization meant that the infrastructure, in terms of the number of servers required for recovery was very large, which posed its own problems. Company Y outsourced the majority of its IT management to a third-party service provider and, due to the lack of specialized equipment being required, decided to outsource its disaster recovery facility provision also. This also suited the geographic diversity of the company. To eventually cover all sites with company-owned recovery facilities would have been unjustifiably costly given that a viable alternative existed.

Company Y did not embark on an extensive analysis exercise in the beginning. The company was well organized and those responsible for the project were able to largely guess the most critical systems from conducting an informal survey across senior management. The responsibility fell to the global IT security group within IT architecture. As these people looked after the strategy relating to the IT infrastructure and were well versed in the operation of the systems, they already had a great deal of intuitive knowledge about what was critical to the business and what was not. When you manage systems and things go wrong you soon find out which systems the business cannot do without. Criticality can often be related to the number of support calls you receive when a system is unavailable, and the seniority and general demeanor of the people calling. This is not very scientific but for a tactical solution where time is important and you have a severe continuity vulnerability to deal with, it is often a sound approach. The key is to continue on with your efforts after this tactical solution is delivered.

Company Y produced a simple RFP that was in essence a shopping list for hardware in a hosted platform along with a set of required response times and so on. They distributed it to companies offering such hosted disaster recovery services and a range of other companies that they had already dealt with for IT-related services. Provision for a basic level of workplace recovery was also made for key technical staff.

Internally, Company Y implemented software for the management of contingency plans and developed these plans in house under the direction of a small number of specialist consultants.

Company Y selected to pursue a proposal from a telecommunications network provider that provided a framework agreement to subcontract specialist disaster recovery providers in any location. This allowed Company Y the flexibility to select the best company in each of the current and future locations where a capability would be required. It also provided a simple working agreement in terms of clear responsibilities, communication, and legalities.

The base capability was created and tested. The project went from inception to completion in around 3 months total elapsed time. The capability was equivalent to that delivered for Company X but cost under $1 million per year on a rolling contract with initial project costs of around $2 million. The creation of the capability for Company X ran over 2-years elapsed time. Company Y is considerably larger that Company X but is also significantly less complex in its IT architecture.

Company Y is now working on long-term strategy for disaster recovery across all facets of the organization with consultants defining requirements in all areas of the organization. Company Y gets away to some degree with driving business continuity from the IT section of the organization simply because the wider organization is generally well ordered and well controlled. This will not last forever though and less advanced organizations would have difficulty achieving the same results. There is already a movement within Company Y to move IT disaster recovery out into a separate business continuity department that is yet to be formed. Company Y is moving through a process of maturity in this area and will almost certainly arrive at a similar level to that defined at the pinnacle of the continuity assurance maturity model.

15

Summary

Firstly, if you have got this far, I would like to thank you for the time you have given to reading this work. You have taken the first step to improving continuity across your organization.

It should be noted that although this is a new methodology, the key components or constituent parts are proven in businesses around the globe. This work simply binds these approaches and technologies together in a practical and pragmatic framework.

Business continuity capabilities are not simple to implement and often require fundamental change across your organization. Because of the vast nature of possible measures that can be taken to reduce risk and assure continuity the job can often seem too large to handle. This methodology aims to break the tasks down into a manageable and straightforward framework that can be followed from start to finish.

There are many projects that you will need to embark on in the implementation of continuity assurance capabilities. I have specifically not discussed project management, as this was not the subject of this work. There is a considerable body of knowledge in existence for project management that can be leveraged and applied to the implementation of continuity assurance.

Today we all live in a dangerous world, where disaster does not simply occur as a result of random events but can be planned by some to be thrust upon others. Disaster is not just natural and indiscriminate but also can be orchestrated and targeted. We, too, must orchestrate our response and target our defenses to maintain our safety, security, and the overall continuity of our way of life.

In the world today our way of life is driven by business. It affects where we live, how we spend our time, how well we live, and how enriched our lives become. Business drives us to create inventions and products with which to fulfill the needs of the world. It is our mechanism for the advancement of our global community and our individual selves. It is a way of life; it is a culture. In protecting our businesses and government agencies we ensure power to our homes, food on our tables, and education to our children. In assuring the continuity of companies and governments alike, we assure their continued operation and the continued pleasure we draw from the finances, products, and services that provide for our continued safety, prosperity, and advancement.

Acronyms and Abbreviations

ACL	Access control list
BAU	Business-as-usual
BCM	Business continuity management
BIA	Business impact assessment
CAAR	Continuity assurance achievement rating
CAM	Continuity assurance methodology
CBRN	Chemical, biological, radiological, and nuclear
CCTV	Closed-circuit television
CDR	Call data record
CEO	Chief executive officer
CFO	Chief financial officer
CI	Configuration item
CM	Configuration management
CMT	Crisis management team
COE	Common operating environment
DAS	Direct attached storage
DR	Disaster recovery

DRP	Disaster recovery plan
EAI	Enterprise application integration
EAP	Emergency action plan
EFTPOS	Electronic funds transfer at point of sale
FTP	File Transfer Protocol
HA	High availability
HR	Human resources
ID	Identification
IDE	Integrated development environment
ISO	International standards organization
IT	Information technology
KPI	Key performance indicator
LOI	Letter of intent
MEA	Multithreaded execution architecture
OQP	Optimum quality point
OS	Operating system
PIN	Personal identification number
PMO	Project management office
QA	Quality assurance
RFI	Request for information
RFP	Request for proposal
RFT	Request for tender
RPO	Recovery point objective
RTO	Recovery time objective
SAN	Storage area network
SCA	Systems criticality assessment
SCM	Software configuration management

SPOF	Single point of failure
SQA	System quality assurance
TDA	Technical design authority
TQM	Total quality management
UHF	Ultra high frequency
UPS	Uninterruptible power supply
VHF	Very high frequency

About the Author

Andrew McCrackan has worked as an independent management consultant for the last 10 years. He specializes in business continuity management, large-scale project management, and change management. He is the founder of Continuity Assurance International. He has consulted to, at one time or another, the majority of the world's largest management consulting firms, and has lived and worked in 11 countries across Europe, the Middle East, Asia, and Oceania.

Born in Adelaide, Australia, he attended the University of Adelaide, under scholarship, where he studied mechanical engineering for a time until getting a job as a software designer and programmer, working on billing and time management systems. He continues to live and work around the world with his wife, Claire-Louise, and their four children. The author can be reached at andrew. mccracken@continuityassurance.com, or http://www.continuityassurance.com.

Index

Recent Titles in the Artech House Technology Management and Professional Development Library

Bruce Elbert, Series Editor

For further information on these and other Artech House titles, including previously considered out-of-print books now available through our In-Print-Forever® (IPF®) program, contact:

Artech House
685 Canton Street
Norwood, MA 02062
Phone: 781-769-9750
Fax: 781-769-6334
e-mail: artech@artechhouse.com

Artech House
46 Gillingham Street
London SW1V 1AH UK
Phone: +44 (0)20 7596-8750
Fax: +44 (0)20 7630-0166
e-mail: artech-uk@artechhouse.com

Find us on the World Wide Web at:
www.artechhouse.com